INTERNATIONAL SERIES IN
NATURAL PHILOSOPHY

GENERAL EDITOR D. TER HAAR

VOLUME 95

Interacting Binary Stars

INTERNATIONAL SERIES IN
NATURAL PHILOSOPHY
VOLUME 95

GENERAL EDITOR: D. TER HAAR

Other Titles of Interest

BATTEN:
Binary and Multiple Systems of Stars

ELGAROY:
Solar Noise Storms

GLASBY:
The Nebular Variables

HEY:
The Radio Universe, 2nd edition

PAPADOPOULOS:
Photographic Atlas of the Stars (3 vols)

REDDISH:
Stellar Formation

STROHMEIER:
Variable Stars

ZHELEZNYAKOV:
Radioemission of the Sun and Planets

NOTICE TO READERS

Dear Reader

If your library is not already a standing order customer or subscriber to this series, may we recommend that you place a standing or subscription order to receive immediately upon publication all new issues and volumes published in this valuable series. Should you find that these volumes no longer serve your needs your order can be cancelled at any time without notice.

The Editors and the Publisher will be glad to receive suggestions or outlines of suitable titles, reviews or symposia for consideration for rapid publication in this series.

ROBERT MAXWELL
Publisher at Pergamon Press

Interacting Binary Stars

JORGE SAHADE

Instituto de Astronomía y Física del Espacio
Member of the Carrera del Investigador Científico, CONICET
Buenos Aires, Argentina

FRANK BRADSHAW WOOD

Rosemary Hill Observatory
Department of Physics and Astronomy
University of Florida
Gainesville, Florida, U.S.A.

PERGAMON PRESS

OXFORD · NEW YORK · TORONTO · SYDNEY · PARIS · FRANKFURT

U.K.	Pergamon Press Ltd., Headington Hill Hall, Oxford OX3 0BW, England
U.S.A.	Pergamon Press Inc., Maxwell House, Fairview Park, Elmsford, New York 10523, U.S.A.
CANADA	Pergamon of Canada Ltd., 75 The East Mall, Toronto, Ontario, Canada
AUSTRALIA	Pergamon Press (Aust.) Pty. Ltd., 19a Boundary Street, Rushcutters Bay, N.S.W. 2011, Australia
FRANCE	Pergamon Press SARL, 24 rue des Ecoles, 75240 Paris, Cedex 05, France
FEDERAL REPUBLIC OF GERMANY	Pergamon Press GmbH, 6242 Kronberg-Taunus, Pferdstrasse 1, Federal Republic of Germany

First edition 1978

British Library Cataloguing in Publication Data

Sahade, Jorge
Interacting binary stars.
1. Stars, Double
I. Title II. Wood, Frank Bradshaw
523.8'41   QB821   78-40301
ISBN 0-08-021656-0 Hard Cover

In order to make this volume available as economically and as rapidly as possible the authors' typescripts have been reproduced in their original forms. This method unfortunately has its typographical limitations but it is hoped that they in no way distract the reader.

*Printed in Great Britain by William Clowes & Sons Limited
London, Beccles and Colchester*

To the memory of

OTTO STRUVE
spectroscopist

RAYMOND SMITH DUGAN
photometrist

HENRY NORRIS RUSSELL
theoretician

CONTENTS

Contents   ix

FOREWORD

The present book grew from an idea that both
of us discussed in 1968 during the "Mass Loss
from Stars" Colloquium in Trieste. Then we
learned that our friend Alan Batten had been
asked to write a book on the subject and we de-
cided to postpone the writing for a few years
in order that our work will complement, rather
than compete with, such an authoritative book as
Dr. Batten's.

In the present book we deal with "interacting
binary stars", that is, with physical systems
that are so close together in space that the
evolutionary history of each of the two components at some stage begins to depart appreciably
from the evolution of single stars.

Binaries are the only source of our knowledge
of the physical parameters of the stars. Actually close binaries - and more precisely eclipsing binaries - are the objects that can provide direct information both on the masses and
the radii of the stars. In a few cases, they
even provide information about the degree of
central condensation and hence the internal
structure.

Now the question arises as to whether the
physical parameters derived from close binaries
are valid for single stars. The definition of
close binaries as given above implied that the
evolution of the components of those systems is
different from that of single stars. Observations indicate strongly that this is so. In fact,
among components of close binaries we find bizarre objects that have no counterpart among
single stars. We could mention β Lyrae, the
eruptive variables, the symbiotic stars, and
many other examples.

Therefore, the masses of evolved components
of close binaries are not necessarily the same
as those of single stars that occupy the same
position in the Hertzsprung-Russell diagram and,
consequently, the physical parameters derived
from close binaries may not be valid for single
stars unless we are dealing with unevolved systems, with one or both components on the main
sequence or above it but before any evolutionary
interaction has already started. Our book will
deal then with an exciting group of objects that
form perhaps over half of our stellar population
and present problems of their own.

The book was written because we felt the need
to collect in one volume the more exciting recent developments together with a brief description of the "classical" ideas from which many of
them developed.

At the Cambridge 1975 IAU Symposium on Close
Binaries one of us stated that the field of
research that deals with close binaries has
"suffered" a "constructive explosion" during the
last twenty years. The discovery that most, if
not all, eruptive variables are close binaries,
the discovery of rapid flickering in the light
of some, the discovery that X-ray sources and
intermittent radio sources are found among close
pairs, a tremendous flux of information indicating the presence of circumstellar material,
theoretical treatments which indicate in some
cases major evolutionary changes in time scales
short even in terms of human history - these are
but a few of the developments which have transformed what was once a rather narrow field of
interest only to a few specialists into one
which touches many branches of modern astrophysics and is engaging in the present tremendous
research activity a large number of scientists
that have come and are coming in from other
fields and branches of science.

It seems proper, then, that in writing this
book we have tried to bring together information
that is scattered over many publications and
papers and to give the present state of the art
with the relevant historical background. We
have tried to give a fair account of the facts
and of the situation in each case and naturally
each one of us has written his part with only
the bias that arises from his own background and
line of activity. We do not pretend to have
exhausted the relevant literature nor to have
done complete justice to all the contributions
in the field. We apologize to those who feel
that we should have mentioned them.

A work like this could not have been carried
without the help of many people and friends who
read parts of the manuscript and offered many
important suggestions and in many cases called
our attention to research that we should not
overlook. We would like to thank for all this
A. H. Batten, K-Y Chen, D. S. Hall, R. H. Méndez,
J. E. Merrill, N. L. Markworth, V. Niemelä, J.
B. Rafert, E. C. Olson, A. Ringuelet, R. E.
Wilson and, particularly, David R. Florkowski,
who went thoroughly and relentlessly with a
critical mind through the whole of the book.

Thanks are also due Jeanne Kerrick who with
admirable devotion, enthusiasm and sense of re-
sponsibility carefully prepared the typescript
and to Elizabeth P. Wood, Adela Ringuelet, Gloria
Nasser, Carlos Hernández, Jorge Wilchepol, Hans
Schrader and Woodrow Richardson who in different
ways helped us during the process of preparing
the book, the latter three with the illustrations.

We acknowledge the help of the Fullbright
Committee, the Consejo Nacional de Investi-
gaciones Científicas y Técnicas of Argentina,
the Instituto de Astronomía y Física del Espacio,
Buenos Aires, and the University of Florida at
Gainesville, Florida, for giving us the opportunity and the facilities that permitted our
work. One of us worked under a Faculty Development Grant from the University of Florida.

Finally, our deep appreciation to our respective wives for their encouragement while the
book was being written.

JORGE SAHADE

FRANK BRADSHAW WOOD

CHAPTER 1

HISTORY OF THE FIELD

1. GENERAL REMARKS

As the title foretells and as stressed in the foreword, we are going to deal in this book with interacting binaries, that is, with physical systems of two stars that are so close together in space that the evolutionary history of each of the two components departs appreciably from the evolution of single stars that, when on the main sequence, occupied the same position in the H-R diagram. We shall feel satisfied with this definition at least for the time being, without entering into the problem as to how a close binary is born.

Thus defined, the term "interacting binaries" becomes more precise than the normally used expression of "close binaries" which was usually defined as a system of two stars, whose separation is of the same order of magnitude as the radii of the objects involved. Plavec (1967) and Paczyński (1967) were the first to suggest, independently, at the same meeting, a change in the way a close binary is defined by taking the evolutionary history into account.

In the development of the field the most heavily emphasized aspects of the research have been different at different times. We recognize four epochs, each one brought about by particular pieces of research or changes of emphasis that actually were breakthroughs in our knowledge in the field. In this introductory chapter we shall go through the four epochs and indicate the highlights of each one of them.

2. EPOCH NO. 1, 1783-1912

For many years, the study of visual binary stars was one of the major fields of astronomical research. The determination of combined stellar masses of the components when an orbit could be computed and a parallax measured, and even individual masses when motions relative to background stars were possible, seemed to be - and was - an achievement of considerable magnitude, even though no one at the time could realize the importance of the knowledge of stellar masses in relation to the

entire evolutionary history of a star. Yet even during the great
nineteenth century era of positional astronomy, close binaries - at
least those that were eclipsing pairs - were not entirely neglected.

This field, both in observation and theory, can be considered as
being started by John Goodricke (1783) with his paper in the Philosophical Transactions of the Royal Society, London. There Goodricke
announced his independent discovery of the variations in brightness
of Algol, although this had previously been detected more than a century before by Montanari and Miraldi, as Goodricke was careful to
point out.   Actually, in old Chinese writings, there is indication
that ancient people observed this variation and the name itself suggests that it was known to the Arabs, but this information of course
was quite unavailable to Goodricke.

Not only did Goodricke independently discover the variability, but
he also found that it was periodic and determined the period. He
published his individual observations (which he had verified by a
friend) to make them available for general use, and gave two possible
explanations for the light changes. One of these - the periodic passing between us and the star of a dark object - eventually proved to
be essentially correct, although more than a century would elapse before this general explanation was confirmed by the variations in
radial velocity found in another eclipsing system.

From Goodricke's time until the early twentieth century, the observations of eclipsing binaries were of the same pattern as those for
other variable stars. The technique at first was by visual estimates
and observers were few, although, well before the twentieth century,
visual photometers both with and without "artificial stars", came into use at a few observatories. Only toward the end of this first
epoch, which we can mark as running from 1783-1912, did techniques
such as photography, spectroscopy, and, starting about 1910, the primitive use of photoelectric effects, began to bring impact from the
developing science of astrophysics. A few astronomers - e.g., Dugan
(1911) and Roberts (1908) - had tried solutions for individual light
curves, but the relations between light changes and the orbital and
eclipse elements were so complex that it seemed no general solution
was possible.

One matter of considerable importance was the discovery of the
"reflection effect" made independently by Dugan (1908) and by Stebbins
(1911).

Near the end of this first epoch the Canadian astronomer Barr (1908)
had made an analysis of the available orbits and velocity curves of
spectroscopic binaries, the first one of which had been discovered
in 1889, and found, (a) that very often the velocity curves were un-
symmetrical, (b) that these asymmetries, with only a few exceptions,
were similar, in the sense that the "ascending branch of the curve"
was of greater length than the "descending branch" and (c) a logical
consequence of (a) and (b) that the values of the longitude of peri-
astron did not show a random distribution but rather were concentrated
in the first quadrant. This peculiar distribution of the curves, known
as the Barr effect in spectroscopic binaries, already suggested as
Barr himself pointed out, that the "observed radial velocities" were
contaminated by some extraneous effect that, at the time, was difficult
to assess.

The problems brought out by the spectrographic observations in this

first epoch were not confined to the Barr effect. As early as 1896
Pickering (1896) reported the discovery by Bailey of variations in
the line intensities of μ^1 Scorpii, a phenomenon which later was found
also in other objects and which appears to be transient in some cases.
The phenomenon can be described by stating that the lines of one of
the components are weaker when its velocities are of recession. Here
again no explanation was possible.

3.   EPOCH NO. 2, 1912-1941

If one event can be said to mark the ending of one epoch - that
which we have called the first epoch in the history of close binaries -
and the beginning of another, it was the publication in 1912 by Henry
Norris Russell (1912a) of the first of his classical papers describing
means for a general solution of light curves. This paper, or some of
the various modifications of it, was to serve as the basis for interpretation of light curves for well over half a century until the avail-
ibility of the electronic computer permitted the use of techniques impossible to apply in practice without these aids. As examples of the
various alternatives, see Fetlaar (1923), Schaube (1924), Sitterly
(1930), Krat (1934, 1936), Piotrowski (1937) and Kopal (1942).

The early paper by Russell, plus the following ones by Russell
(1912b, 1912c) and by Russell and Shapley (1912a, 1912b), still deserve
care il reading by the serious student of close binary systems. As-
sumptious were made which were entirely justified in the light of the
precision of the observations then available, but which were not entirely satisfactory in interpreting the more precise observations that
result with the widespread use of photoelectric photometry that followed the development of the multiplier photocell some thirty years
later. Further it is clear that this approach was never intended to
apply to systems in which the components were so close that the interaction effects between the components caused large changes in the observed brightness. This will be discussed in more detail in Chapter
3.

These papers came at a crucial time. The advent of photography had
permitted the discovery and identification of eclipsing systems at a
far faster rate than ever before. In the years immediately following
larger telescopes and more spectrographs were built so that more radial
velocity curves were to be observed. Elements from these combined with
the photometric elements, of course, gave the absolute dimensions of
the components. By 1922, this had been done for 14 systems (Plaskett,
1922). Not only were more light curves being observed, but in the
years immediately preceding and following, more precise photometric
observing techniques were developed. Increasing use of the polarizing
photometer and the wedge photometer both gave measures which not only
were more precise than estimates, but which were essentially free from
the systematic errors which plagued both visual and photographic estimates. In either of these instruments, the image of the variable was
placed by optical means in a horizontal line with and at a convenient
distance from either the image of an artificial star or that of a
nearby comparison star. The brightness of one image was varied by the
use of an optical wedge or a Nicol prism until the images were equally
bright. The human eye is at its best when estimating when two sources
have nearly identical luminosities. Tests have shown that, as long as

significant color differences are avoided, different observers get
the same results to within observational scatter. The significance
of this lies in the fact that, if light curves observed with these
techniques show significant differences at different epochs, the
possibility of the changes being real and in the stars themselves
must be seriously considered.

Photoelectric techniques also improved and became somewhat more
common although even in the late 1930's there were only eight observatories in the world at which working photoelectric photometers
existed and, on any given night, probably at least four of these were
inoperative because of technical difficulties. Further, the sensitivity was such that observations were confined to the brighter
stars and were only in the natural color system of the telescope-
photocell combination.

The highlights of the thirty years following Russell's paper can
be briefly summarized. Shapley (1915) applied these methods to the
existing light curves. Most of these were by visual or photographic
estimates, and doubtless contained systematic errors especially when
the primary eclipses were deep. A long series of observational papers
by R. S. Dugan and his students presented light curves of precision
exceeded only by photoelectric methods, and solutions and interpretations on the "Russell model". Techniques of making photographic
measures as distinct from estimates (e.g., photoelectric measures
of the densities of out-of-focus photographic images) were developed,
but their applications to eclipsing systems were relatively few. However, the epoch was essentially one of consolidation - more observers,
more systems discovered, more light and velocity curves, and more
solutions.

Not all of the work necessarily meant progress. For example, in
many cases in the W Ursae Majoris systems, conventional rectification
treatment - i.e., removal from the light curve before proceeding to
the solution, of all effects other than those caused by eclipses -
gave coefficients for the "reflection effect" of the opposite algebraic
sign to that required by the theory. Interpreted by the conventional
model, this implied that the side of the star of lower temperature
which faced that of the hotter star actually had its temperature decreased by the flux of radiation from the high temperature component.
Almost the only - and certainly the most likely - alternative to this
rather startling interpretation was that the simple Russell model, at
least as far as rectification was concerned, was not suitable for
dealing with all systems. The main difficulty seems to have been
failure to recognize the intrusion of "complications" into the rectification formula, but in many cases solutions were forced through,
and the results appeared first in the literature and then in various
uncritical compilations.

One extremely important theoretical development, which for practical reasons was not exploited to its fullest, was the realization that,
if the stars are distorted and moving in elliptical orbits around their
common center of mass, the major axis of the ellipse will rotate at
an angular speed which is determined by various factors, one of which
is the degree of central condensation of the components. In many cases,
this will essentially be determined by one component alone, and thus
offers an observational test of an extremely important stellar property. The original paper calling attention to this was also by Russell
(1928). However, he neglected to allow for the changing shape of the

stars with their varying separation at different parts of the elliptical orbit. Cowling (1938) called attention to this and derived a
correct formula. Later work took into effect certain higher order
terms.

Various writers tried to apply these ideas to individual systems,
often with conflicting results. Three factors contributed to this.
The rate of rotation of the line of apsides (relative to the orbital
period) had to be determined from the observational data in order for
the central condensation to be computed. Since the apsidal period
was usually of the order of many decades, the observational data normally did not have a sufficiently long time lapse to permit accurate
determination, and estimates of this by various authors for the same
system often varied widely. Further, the values of the relative radii
were needed to a high order of precision, and a sound value of the
mass-ratio was required; both of these requirements presented difficulties in many of the cases in which significant apsidal rotation
has been detected. This will be discussed at greater length in Chapter 6.

The photometric work during this interval was characterized chiefly
by the improvement of techniques - the constant deflection method in
photoelectric photometry was perhaps the most significant - and the
production of an increasing number of light curves by various means.
The competition for precision ran high, especially among the photoelectric observers, and strong efforts were made to produce the smallest possible value of the probable error of an individual observation
from a computed curve. Instrumental difficulties were such that unusual variations were rarely attributed to the star itself. "Night
corrections" or "seasonal corrections" were frequently applied to the
observations in order to produce light curves with small scatter, and
the size of these corrections, or even the fact that they had been
applied, was stated only rarely. Almost all astronomers of the time
seemed to feel that light curves should repeat smoothly from cycle
to cycle. In fact, when a good curve had been observed and a solution
made, it was often labeled "definitive" and it was felt that no significant amount of information would be obtained by additional observations. The idea was strengthened when "humps" on light curve reported by early observers were not duplicated in later observations.

The net effect was to produce light curves and solutions of varying
degrees of determinancy; combinations of these with results from velocity curves gave absolute dimensions of varying degrees of reliability.
However, it should be remembered that this was the only general way
we could generate our knowledge of stellar masses, radii, and densities.

Even during this time, evidence was accumulating which suggested
the model was oversimplified, and not all astronomers accepted it
blindly. As early as 1898, Pickering (1904) and Wendell (1909) independently observed a strange increase of brightness immediately
after primary on the light curve of R Canis Majoris. Dugan (1924)
observing a number of years later, could find no trace of it in his
own observations, but after a careful study of the earlier work concluded that it was a real but transient feature of the light curve.
A series of velocity curves by Jordan (1916) showed distortions indicating an eccentric orbit, in complete contradiction to the light
curve, but with an "eccentricity" which varied from year to year.
Wood's (1940) photoelectric photometry confirmed the circular orbit

indicated by earlier light curves but gave depths of minima which could not be reconciled with those found by Dugan. Dugan had already called attention to a discrepancy between the depth of secondary minimum he found as compared to that in Wendell's light curve. He concluded the difference was significant, but could at that time offer no explanation.

The puzzles were not confined to one system, nor were they suggested by the photometry only. As early as 1922, Shapley (1922) noted a difference in the appearance of the ionized calcium lines on the leading and following sides of the brighter component of β Aurigae. Struve (1935) discussed the asymmetry of spectral lines in a few eclipsing variables and more were discovered that displayed the type of line intensity variations that had been discovered by Bailey. Lause (1938) suggested a change in diameter of at least one component of an eclipsing binary.

This is not a complete listing, but if we compare it to the vast number of papers assuming complete stability it would seem as though there was ground for the firmly established concept of close doubles as highly stable systems constantly repeating the same light and spectrum changes. A few notable examples to the contrary should be mentioned briefly.

Carpenter (1930) observed a velocity curve of U Cephei indicating a highly eccentric orbit. Yet Dugan's (1920) light curve showed a centrally located secondary of the same width as the primary, almost certainly an indication of a circular orbit. The discrepancy was real; there was no easy way to rationalize it, and none was attempted.

Dugan and Wright (1937) presented the first general study of period changes in eclipsing systems. A few attempts had been made earlier of individual systems (e.g., Algol), but this was the first time in which a number of systems were studied together. This came at the time when the idea of apsidal rotation was creating some interest. Rotation of apse would cause a sinusoidal variation of the (O-C) residuals of times of minima in both primary and secondary eclipses with the curves being 180° out of phase, while light time effects caused by motion of the system around a third body would produce periodic changes affecting both minima equally. However, in almost every case, Dugan and Wright found relatively sudden changes with no periodic pattern; the changes could increase or decrease the periods with no clear preference shown. Again no theoretical explanation could be offered.

There were other spectrographic results that pointed also to the fact that things were perhaps not as simple as we might have liked. We could mention the very small mass function yielded by systems like R Canis Majoris, the meaning of which was only clear several years later. Furthermore, Wyse (1934) had called attention to the fact that the presence of H in emission, and in some cases of Ca II in emission, was a frequent occurrence in close binaries. There was the very peculiar spectrum of β Lyrae, a system that had attracted the interest of astronomers ever since its variability was discovered, by Goodricke (1785) in 1785, and which had defied interpretation in spite of the several attempts to analyze and understand it. The first investigation of the spectrum was due to Belopolsky (1893) who found that the radial velocities from the Hβ emission gave a velocity curve opposite in phase to and with smaller amplitude than that from the stellar absorption lines. It is remarkable that a few years later, Myers (1898) proposed the existence of a gaseous envelope around the system of β Lyrae. In fact, β Lyrae, together with ε Aurigae, posed an additional and very challenging puzzle, namely, the fact that both systems have deep primary eclipses and yet no spectral lines of the secondary components are ever seen, while the lines of the primaries are present at all times.

Thus, notwithstanding the fact that , in general, astronomers believed that close binaries were uncomplicated systems of two stars and tended to expect that their velocity curves should reflect purely orbital motions and to believe that discordant velocities had to be discarded, there emerged observational facts that clearly indicated that there were quite a number of problems which needed special attention.

4. EPOCH NO. 3, 1941-1966

If the change to the second epoch was triggered by a theoretical paper, the change from the second to the third was largely due to a rapid series of observational data by Otto Struve and his collaborators in the spectrographic field. This has appropriately been called the "Struve Revolution" (Popper, 1970). Eventually it altered profoundly our whole concept of the stability of close binaries. Almost simultaneously, the multiplier photocell made a major change in photometric techniques.

The ground for the "Struve Revolution" had been set by the growing realization that close binaries were not systems as simple as astronomers had previously thought and that, as we have previously pointed out, there were already a number of systems known that posed questions which required answers. Clearly this, in turn, required a systematic and massive investigation, concentrated in time, of a large number of systems, particularly of those that display some kind of peculiarities in their spectrum and those whose elements as suggested by the photometry were different from those indicated by the spectrographic observations.

This approach had a start at the end of our second epoch with the study of ε Aurigae (Kuiper, Struve, and Strömgren, 1937) made at the Yerkes Observatory. It was followed a few years later by an investigation of β Lyrae (Struve, 1941; Kuiper, 1941; Gill, 1941; Greenstein and Page, 1941) with the participation of several distinguished astronomers, each of whom attacked different aspects of the problem. This investigation led Struve to postulate for the first time in close binaries the existence of gaseous streams between the components and of an outer expanding general gaseous envelope to interpret some of the spectral features that were displayed by the star. The understanding that was reached of this peculiar system actually represented a tremendous breakthrough in the understanding of the complexity of the structure of an interacting binary. The outcome would have been even more far reaching were it not for the fact that in the early nineteen forties the belief was firm that every star had to obey the theoretical mass-luminosity relation as formulated by Eddington.

The success that had been attained, the further discovery by Joy (1942) that the double H emissions displayed by the eclipsing system RW Tauri undergo eclipses as well and therefore must arise from a gaseous ring around the primary component, and moreover, the fact that

during World War II Struve had a large amount of observing time at
his disposal at the McDonald Observatory, led him to undertake, with
collaboration from Hiltner, Cesco, Sahade, Hardie, and others, the
intensive study of a number of systems. This trend, was also followed
somewhat less intensively by other scientists, who later continued
and extended the avenues that had been opened.

The strong effort, that started in the early forties and had impetus
over some twenty years, led for the first time to an understanding of the
Barr effect in terms of the distortion of the velocity curves of close
binaries whose gaseous matter streams out from one of the components
of the system, normally from the less massive star towards the other.
This phenomenon can lead to the formation of gaseous rings or envelopes around the primaries and eventually of expanding envelopes that
surround the entire systems. In this period the so-called Algol paradox, (the star that had apparently evolved farther was the less massive component) was evident. An explanation was suggested by Crawford
(1955) and by Kopal (1955) in terms of post main sequence evolution.
A decade later, through the detailed computation of models of close
binary evolution, it was shown quantitatively that this process could
account for many of the observed features. Whether it is the only
one involved is a matter for future work.

Struve's insight and vast astronomical knowledge made the field of
close binaries a lively and exciting realm of research. Nevertheless,
until the end of this second period the number of scientists engaged
in the study of close binaries was rather small and it was even difficult to induce other astronomers to engage in the study of specific
objects. The situation now is completely the opposite.

By the middle fifties our understanding of the evolution of single
stars was taking shape and it then became possible to start thinking
about peculiar close binaries in terms of their evolutionary history.
It was then that Sahade (1958; Sahade et al. , 1959) placed emphasis
on the secondary component of β Lyrae probably being the more massive
component of the system, a conclusion that was generally accepted
later on various grounds. Actually this conclusion, had been suggested by Belopolsky (1893) and Curtis (1912) many decades earlier
on the basis of the radial velocity curves derived, as we shall see
in Chapter 10.

We should mention as additional highlights of this third epoch,
one extremely rich in results that changed drastically the classical
view of double star astronomy, and set the stage for further and more
challenging developments, the following examples:

(a) The conclusion, through the study of systems like XZ Sagittarii (Sahade, 1945, 1949) that the values of the mass-ratio in close
binaries could be quite large, at variance with previous beliefs that
it should be around one.

(b) Wood's (1950) recognition that the period variations that
characterize a number of eclipsing systems probably are indicative
of mass loss.

(c) The discovery of the binary nature of the recurrent nova T
Coronae Borealis (Sanford, 1949), of the explosive variables SS Cygni
(Joy, 1954) and AE Aquarii (Joy, 1956), of the Nova DQ Herculis (Walker,
1954) and the subsequent study of a number of old novae (Kraft, 1964)
which suggested that all eruptive variables with the possible exception of supernovae are binary objects.

(d) Huang's (1963, 1965) suggestion of the existence of a thick, flat envelope, which he called "disk", around the secondary components in ε Aurigae and β Lyrae to explain the apparent incoherence that results from the fact that the spectra of the primary components are seen at all times and the secondary spectra are not detected even at the center of the primary eclipses in spite of their depths, as we have previously stated.

The exciting spectrographic developments were accompanied by developments in photometric theory and especially in the observational field. Initially, and for the most of the following twenty years, the theoretical advances were essentially refinements of practical applications of the Russell model. Russell (1942, 1945, 1946, 1948a, 1948b) published a series of papers dealing with methods of solution and giving a treatment of rectification which took into account many of the developments which had occurred since his initial paper. Krat (1941), and Kopal (1947) published sets of auxiliary tables. Irwin (1947) published tables designed to facilitate a least squares solution.

A major contribution in this area, was the publication by Merrill (1950, 1953a) of an extensive set of tables of the Russell functions for the solution of light curves of systems having components with limb darkening equal to 0, 0.2, 0.4, 0.6, 0.8, and 1.0. A list of historical and general references was included.

These tables were contemporaneous with a set of nomographs for the rapid solution of light curves (Merrill, 1953b) and with a complete discussion of and methodology for solutions on the Russell model including rectification on that model (Russell and Merrill, 1952).

These were by no means the only developments in theory or in its application. Other efforts, too numerous to list here, were published As examples only, we could mention Hosokawa's (1955) tables of darkening coefficients, Ovenden's (1956) paper on the photometric effects of gaseous envelopes, Kitamura's (1959) compilation of elements of W Ursae Majoris systems, Plaut's (1953) compilation of elements of all systems brighter than apparent photographic magnitude 8.5, Linnell's (1958) treatment of atmospheric eclipses, Grygar's (1962) consideration of non-linear laws of limb-darkening - these offer but a few samples of the tremendous amount of work produced. The general references in the Card Catalogue of Eclipsing Binaries now kept at Florida lists approximately 500 papers published in this (1941-66) interval on topics in the field; this does not include work on individual systems which are listed separately.

There were a few indications on the changes that were to come beginning around 1966, or possibly the years immediately preceding. Kuiper (1941) had invoked the "Roche" model in his investigation of β Lyrae and discussed the significance of the outer Lagrangian point. Wood (1946) had used it in arriving at a limiting solution for R Canis Majoris, and had again (Wood, 1950) called attention to its utility in a discussion of abrupt period changes. His suggestion that for practical purposes, close double star systems could be divided into two classes, one having both systems removed from the inner contact equipotential surface and the other having at least one component close to it was later extended somewhat by Kopal (1955) who suggested dividing the latter class into two groups - one having one component only near the limit, and the other somewhat controversial one consisting of two stars each exactly filling the critical lobe. Plavec (1964) computed tables for this model. Various authors (e.g., Plavec and

Kríz, 1965) computed trajectories of ejected particles in close binaries. In theory, too, the time for change of emphasis was approaching.

Photometric observations, also, were calling for different concepts.
The development of the multiplier photocell, (plus the rarely mentioned, but almost equally important general developments in the field
of commercial electronics) gave far greater reliability and sensitivity (amplification within the tube by a factor of 10^6 instead of the
usual factor of about 30 when only gas amplification was possible)
and thus brought within range of telescopes of moderate size systems
previously too faint for any but the largest instruments. A further
important factor was that an astronomer could now make photoelectric
observations without either being an expert in electronics or a member of a team containing such experts. The result was both a large
increase in the number of photoelectric observations produced, and
beginning in the early 1950's, an increasing use of color filters. A
summary of the instrumentation to 1953 and some of the developments
in the years immediately following may be found in the proceedings
of two symposia (Wood, 1953; 1957a).

The effect of these developments on the field of photometric observation was not surprisingly a stimulating one. The evidence for
photometric instability until 1955 has been briefly summarized by
Wood (1957b). An excellent discussion of the developments in all
fields (spectroscopy, photometry, and theory) in this interval has
been given by Plavec (1968).

In general, it was clear that photometry also, now of the precise
photoelectric type whose accuracy was limited chiefly by the earth's
atmosphere, was giving increasing evidence that much more than simple
eclipse theory and the conventional "proximity effects" were needed
for the interpretation of light curves, and that not only were relatively short period changes in the light curves frequent in some
types of systems, but further that some systems (AO Cassiopeiae is one
example of many) almost never showed the same curve when observed at
different seasons. The concept of the limiting surface was receiving
increasing attention from observers in interpreting their observations,
and irregular period changes were being more intensively studied.
The original suggestions made in 1950 were expanded by various astronomers into more sophisticated treatments - examples are papers by
Piotrowski (1964) and Kruszewski (1964).

Two papers in particular should be mentioned as indications of
changes to come in the interval beginning roughly in 1966 and still
continuing. The first by Crawford (1955) has already been mentioned;
it appeared in 1955. The theory of post main sequence evolution for
single stars had been developed, in particular for the expansion to
the red giant stage. Crawford suggested that the secondaries of typical Algol systems were originally the more massive components, and
had expanded to the critical limiting surface, and then transferred
much of their mass to the companion.

The second paper by Morton (1960) was of an entirely different
nature, but led to developments which converged with Crawford's ideas.
Morton applied the recent developments in studies of stellar evolution to the field of close binaries in as rigorous a way as was possible at the time, He considered what happened when the more massive
primaries, in their stage of rapid evolution from the main sequence
to the red giant state, filled their limiting lobes and presumably

began transferring mass to the secondaries.

Essentially Morton computed a series of models in thermal equilibrium. He removed an arbitrarily chosen amount of mass and then computed a new radius. For stars of this particular stage of evolution the new radius was found either to be essentially the same or slightly larger than the preceding one. However, the transfer of mass from one star to the other causes the critical zero velocity surface to expand around the secondary but to close in around the primary, thus creating an unstable condition and increasing the rate of mass loss. The total time involved until a stable condition is again reached is extremely short - about 10^5 years for a star of 10 solar masses.

Morton's calculations did not consider change of either period or separation of the components during this process. A little later, Smak (1962) made similar computations assuming conservation of total mass and angular momentum. Little more could be then done until more was known of the evolution and structure of single stars.

5.   EPOCH NO. 4, 1966-

The choice of 1966 to mark the beginning of what may be called the "modern" epoch is somewhat arbitrary, as was the choice of 1941. Nevertheless, it was about that time that major emphasis began in three areas which are being pursued vigorously today. These were: (1) the application of computer techniques to the solution of light curves; (2) the application of new ideas concerning stellar structure and evolution to close binary systems; and (3) the use of satellite telescopes to permit observations in the far ultraviolet and X-ray regions of the spectrum.

The application of computer techniques to the solution of light curves in order to determine from the light changes the orbital and eclipse elements developed two basic approaches. One of these essentially carried on the basic philosophy of the Russell-Merrill treatment. The other dispenses with separate treatment of rectification and attempts to solve for all parameters simultaneiously. These will be more completely discussed in Chapter 3.

These were of course, by no means the only contributions. We can mention but a few examples by way of illustration. Merrill (1970) showed that proper treatment by the conventional rectification method required better knowledge of higher harmonics than had previously been realized. Various authors (e.g., Sobieski, 1965a, 1965b; Rucinski, 1973; Chen and Rhein, 1971, 1973; Ureche, 1972) discussed the reflection effect in more sophisticated treatment than had previously been possible. Tidal distortion was considered in more detail by various authors (e.g., Wilson and Sofia, 1975). Kopal (1976) continued development of the Fourier analysis of light curves; Budding (1974) discussed the application of optimization techniques. Cherepashchuk (1969, 1975) discussed the Wolf-Rayet systems CV Serpentis and V444 Cygni on the basis of narrow band observations. Indeed, more than 75% of Volume 12 of Vistas in Astronomy edited by A. Beer consisted of papers on close double stars. It is impossible to review the entire flow of work in any reasonable space. Selected aspects will be discussed in Chapter 3.

One extremely important development in another direction was the
work of Lucy on W Ursae Majoris systems (e.g., Lucy, 1968, 1973,
1976). Many of these extremely close systems had defied precise
analysis by conventional means and the puzzle of mass-ratios which
were completely inconsistent with relative luminosities remained a
serious one. As previously noted, many years earlier Myers (1898)
had attempted to explain some of the peculiarities of β Lyrae by postulating a common envelope surrounding both components. Lucy adapted
this idea to W Ursae Majoris systems; using a modern astrophysical approach, he assumed a common convective envelope for the pair and while
all problems are by no means solved, this general approach still seems
by far the most promising.

The second major area was the application of newly developed ideas
of stellar structure and evolution to the field to close binaries.
The major impetus came from what has been called the "Algol paradox".
Even in the 1930's specialists in the field were well aware that the
fainter components of these systems were many times too bright for
their mass if compared to the conventional mass-luminosity relation
or, expressed in another manner, were found in the H-R diagram in a
region about the main sequence, now conventionally called the region
of the subgiants. The puzzle had resisted all attempts to solve it.
For example, Wood (1964) suggested a process whereby components of
different chemical composition could form out of the same cloud during
the contraction process. However, this was at the time when little
was known about these stages in single stars, and in considering the
relation of this model to reality, he concluded that a great deal more
work needed to be done before we could accept or reject it. Another
reason for choosing 1966 as the beginning of a new epoch was that in
that year, an international colloquium considering evolution of double
stars was held at Uccle in 1967. Here, it was learned that detailed
calculations of a series of evolutionary models had started nearly
simultaneously at three separate observatories. These were at Göttingen (B. Kippenhahn and collaborators), Warsaw (B. Paczyński and
collaborators), and Ondrejov (M. Plavec and collaborators). Henyey's
method of calculating stellar models was available, and the general
approach of calculating a series of models of constantly decreasing
mass to find the effect on the mass losing star was followed.

These were successful in tracing the evolution on the H-R diagram
of the more massive star from the main sequence to the subgiant region
and in at least one case, to the white dwarf stage. Two general classifications were first suggested. In each, the more massive star evol-
ves more rapidly and essentially as a single star until it reached the
limiting lobe. In case A, it reaches this when its energy is supplied
by hydrogen burning in the core and it is still on or near the main
sequence in slow expansion. In case B, the star is already somewhat
evolved, has a helium core, is burning hydrogen in a surrounding shell
and reaches the critical lobe when it is in a state of rapid expansion
to the red giant stage. Two other possibilities exist. In case C,
the overflow of the liminting lobe occurs when the energy generation
results from the He-burning core and the H-burning shell. Finally, in
case AB, the lobe is filled within the conditions of case A and the
outflow proceeds to the situation depicted in case B. In Chapter 7 we
discuss binary star evolution.

The most recent developments in this field, now underway at several
centers, include the effect of mass addition to the originally less

massive star and its subsequent evolution, and the effects of mass loss to the system. This latter effect is difficult to handle theoretically both because it also involves loss of angular momentum and because the loss may be explosive in the form of eruptive prominences or even more violent ejections. However, it almost certainly corresponds more closely to the real universe than assumptions of zero mass loss from the system.

The third general field of expanded knowledge is of course the observational field. In general, this has kept well ahead of the theory and has frequently presented completely unanticipated phenomena. One important reason, but by no means the only one, was the observation of hitherto unattainable regions of the ultraviolet and of higher energy ranges made possible by observations from satellites. Probably the most dramatic discovery has been that of X-ray binaries, with the interpretation that X-radiation results from the conversion into thermal energy of the kinetic energy of the material being lost by a relatively "normal" component and falling onto a collapsed companion or a dense envelope (disk) surrounding the latter.

It is interesting to note that the systems observable from satellites now reach roughly the same limiting magnitudes as those observable photoelectrically by earth-based telescopes of the same size before the development of the multiplier photocell.

Another development, still in its infancy, is the detection of radio emission at irregular intervals from certain close double star systems. Yet another, is the "flickering" so far studied chiefly by Warner and by Warner and Nather in the case of U Geminorum stars, and explicable by a "hot spot" created by ejected matter falling on a disk surrounding the small component. Incidentally, the concept of the hot spot was developed almost simultaneously by Krzemiński and by Smak from conventional photoelectric observations. Each of these developments will be discussed in more detail in subsequent chapters. Paczyński (1965) has given a summary of the state of knowledge at the beginning of the epoch and Warner (1976) has summarized at least one subfield as of 1975.

The understanding of the nature of the elements that define a close binary and its structure has led to a considerable progress in the interpretation of many hitherto difficult to understand peculiar objects.

Finally, attention should be called to the fact that the tremendous volume of work being carried out in the field of close double stars implies that the preceding section is far from complete, and that it is not only possible but probable that work which will later prove significant has been overlooked. An example of the volume of work is found from an inspection of the previously mentioned Card Catalogue of Eclipsing Binaries now kept at the University of Florida. One section lists only "General References", that is, theoretical papers or summaries not devoted to any particular system or systems. From January 1966 until December 1976, nearly 1600 papers appearing in standard journals or observatory publications are listed.

We shall now consider some of the principal topics in more detail.

CHAPTER 2

THE ZERO-VELOCITY SURFACES

Since most of the discussion and interpretation of interacting binary systems is made by representing the two components surrounded by the so-called "Roche equipotential surfaces" as though these are some sort of fixed, omnipresent, features physically connected with binary systems, we shall devote this Chapter to the review of such an important concept.

To deal with this concept is to deal with the restricted three-body problem. As it is well known, the n-body problem is, in principle, not solvable and only arduous numerical calculations can produce a solution with the needed precision. However, in particular cases the problem can be described by a set of algebraic equations. One such particular case is the restricted three-body problem which considers the motion of an infinitesimal body in the gravitational field of two finite mass-points that describe circular orbits around their center of gravity. By infinitesimal we mean that the particle mass is so small that its attraction over the other masses is negligible. This problem was considered for the first time in a thorough manner by the French mathematician Joseph Louis Lagrange in 1772 (cf. Szebehely, 1967) who found that there are five positions, that are known as the Lagrangian points L_i (i = 1, ...,5), where the motion of the infinitesimal particle, if placed on them, will follow a circular orbit always maintaining a fixed orientation relative to the finite bodies; that is, the particle will remain relatively at rest unless perturbed by external forces. The five Lagrangian points are represented in Fig. 2.1. L_1, L_2 and L_3 are on the axis joining the two mass points and constitute the "straight-line solution" while L_4 and L_5 are so located that they form equilateral triangles with the finite bodies and constitute the "equilateral triangle solutions". L_1 is between the finite masses and L_2 and L_3 are on the "outside", L_3 being at the side of the larger mass. The solutions are all periodic, but if we change slightly the initial conditions (position and velocity of the infinitesimal body) that led to those solutions, the new orbits will deviate rapidly from the periodic ones in the case of L_1, L_2 and L_3 and will never depart much from them in the case of L_4 and L_5 if $\mathfrak{M}_1/(\mathfrak{M}_1 + \mathfrak{M}_2) < 0.0385$, where $\mathfrak{M}_1$ and $\mathfrak{M}_2$ are the masses of the two finite bodies.

In nature, we find that the Sun, Jupiter and the Trojan asteroids conform to the equilateral triangle solutions, while the Gegenschein corresponds to the straight-line solution with the Sun and the Earth.

Now let us turn to the motion of the infinitesimal body. Let us consider it in the framework of the rectangular coordinate system that has its origin at the center of mass defined by the two finite bodies, the x-axis along the line joining them and the z-axis perpendicular to the orbital plane. If we integrate the differential equations of motion we obtain the expression

$$V^2 = x^2 + y^2 + 2(1 - \mu)/r_1 + 2\mu/r_2 - C. \qquad (1)$$

where μ and $1-\mu$ represent the masses of the two finite mass-points in units of the total mass, r_1 and r_2 are the distances of the infinitesimal body from them, in units of the distance between the mass-points, and V is the velocity of the particle.

The constant of integration C is defined by the initial conditions, and if C is known, expression (1) gives the velocity of the particle at any permissible point of the rotating space. In particular, if we make V=0 we define a surface, called the zero relative-velocity surface, at one side of which the velocity values will be real, and, at the other side, imaginary; that is, the particle will be able to move only on one side of that surface. The shape of the surfaces depends on the value of C as one can see in Fig. 2.1 and there are particular ones that intersect precisely at L_1, L_2, and L_3, respectively.

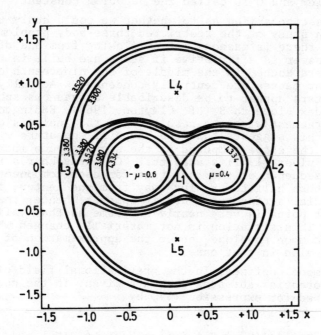

Fig. 2.1.   Zero relative-velocity surfaces and the five Lagrangian points.

For large values of C the zero-velocity surfaces take the form of closed ovals around each of the bodies, at first differing not at all or only slightly from spheres. If we decrease C we come to two ovals that touch each other at L_1 giving rise to a dumb-bell-like configuration. For even smaller values of the integration constant the surfaces embrace the two finite masses and eventually open up at L_2 and then at L_3. The surfaces vanish from the xy-plane at L_4 and L_5.

The surface that contains L_1 is usually called the "inner contact surface" while those that contain L_2 and L_3 are designated as "outer contact surfaces". Kuiper (1941) and Kopal (1954) have computed, for different mass-ratio values, the zero-velocity surfaces passing through L_1, and Kuiper and Johnson (1956) have computed in addition, the zero-velocity surfaces that contain L_2.

The integral expression (1) was found by the German mathematician Karl Gustav Jakob Jacobi and can be written as

$$v^2 = 2U - C \qquad\qquad\qquad (2)$$

where

$$U = (1/2)(x^2 + y^2) + (1 - \mu)/\, r_1 + \mu/r_2 \qquad\qquad (3)$$

Such an integral is usually called the Jacobian integral of the restricted problem and C is called the Jacobian constant.

Now the question arises as to whether we could apply the conclusions from the study of the restricted three-body problem to binary systems where there is gaseous matter flowing from one of the components. The answer is affirmative if stars are built in the model proposed by Edouard Roche, in the middle of the nineteenth century, that is, if the star masses are centrally condensed. Actually, in radiative equilibrium stars appear to be describable as gaseous spheres with polytropic index close to 3 (cf. Clayton, 1968), Eddington's "standard model", in agreement with the results from the analysis of apsidal motions, and this suggests that the shape of the stars will deviate only slightly from the shape computed on the simple Roche model (Chandrasekhar, 1933). Kuiper (1941) states that "Chandrasekhar's result is obviously a consequence of the fact that for each component 90% of the mass is less than half the radius away from the center, in which region the equipotential surfaces are nearly spheres, hence, the attraction on an exterior point is very nearly the same as that valid for the Roche model. This situation is not materially changed when the components have a common envelope; hence the approximation of the Roche model is good also in that case".

The Roche model defines, in the gravitational field of a binary system, equipotential surfaces that are given, in the coordinate system described for expression (1), by

$$- \Omega = (1/2)(x^2 + y^2) + (1 - \mu)/r_1 + \mu/r_2 \qquad\qquad (4)$$

where Ω is the potential.

Expressions (2) and (4) yield

$$V^2 = -2\Omega - C \qquad\qquad (5)$$

Therefore, the equipotential surfaces corresponding to the Roche model
coincide with the Jacobi's zero-velocity surfaces. Thus we talk of
critical equipotential surfaces when we refer to those that contain
L_1 or L_2 or L_3.

In the Roche model the shape of each component of a detached system, in Kopal's nomenclature, is defined by two equipotential surfaces
that correspond to their actual radii.

Kuiper (1941) was the first who realized the significance of the
Lagrangian points in trying to reach a theoretical understanding of
Struve's (1941) interpretation of the spectrum of β Lyrae, as L_1 could
provide "a route for transferring mass from one star to the other,
while L_2 and L_3 control the flow of mass from the system to the outer
space" (Struve, 1957). As Struve (1957) further pointed out, at L_1,
L_2 and L_3, Ω has maxima along the x-axis, and a minimum in the y and
z directions; "consequently, the three points are the saddle points
with respect to Ω and can be regarded as the mountain passes on a
highway".

Let us at this point summarize the conditions under which the considerations that we have made regarding the application of the restricted three body problem are valid. Those conditions were that:

   a) the components of close binary are stars in radiative
       equilibrium;

   b) the mass distribution approaches the definition of mass-
       points;

   c) the orbital motions are circular,

plus the following two which are taken for granted, namely,

   d) there is synchronization between axial rotation and orbital
       revolution;

   e) the axes of rotation are perpendicular to the orbital plane.

There are now a few attempts to try to see what happens if there
are departures from these conditions or we take other conditions into account.

Plavec (1958) considered departures from condition d) and e). The
results of his study indicated that "L_1 comes the nearer to the center
of gravity of the rapidily rotating component the faster it rotates",
that is, "the critical surface contracts" if we consider a component
that rotates with a velocity higher than that which corresponds to
synchronization. In the case that the axes of rotation form with the
orbital plane an angle different than 90°, "L_1 would periodically
change its position in space and the equipotentials would change their
shape, the more so, the faster the components rotate and the more are
their axes inclined".

In the case of elliptical orbits we have a different set of zero-
velocity curves for every value of the true anomaly (Szebehely, 1967b).
If the orbits are eccentric the expression for the potential is similar
to that for non-synchronous rotation except for a factor that is proportional to the third power of the instantaneous separation of the

two finite masses (Avni, 1976).

A problem that has been taken up during the last few years is the
question of the effect of radiation pressure upon the equipotential
surfaces as discussed above. Schuerman (1972) undertook this question
by considering the radiation pressure from one of the components of
the binary system, something which was accomplished by replacing the
mass of that component by the value of the mass multiplied by a factor
$(1 - \delta)$. δ represents the ratio of the radiation pressure to the
gravitational attraction and reflects the effect of the radiation
field which is assumed to be independent of position. The shadow effects of the companion and the effect of the pressure on the latter
star are neglected. Schuerman finds that the effect of radiation pre-
sure is to modify the "classical" equipotential surfaces and the extent
of the modification depends on the value of δ and the mass-ratio. In
some cases L_2 and L_3 share the same equipotential surface and for
values of δ larger than a critical value δ_c, a contact surface no
longer exists. Schuerman's results have been applied by Kondo and
McCluskey (1976) and by Kondo et al. (1976) to actual systems which
are X-ray binaries.

More recently, the effect of the radiation pressure of the two components of a binary system has been considered by Zorec (1976). The
results were similar in the sense that Zorec's investigation led to
the conclusion that contact surfaces no longer exist for values of
$\delta > \delta_c$ and further suggests that radiation pressure favours the formation of multiple circumstellar shells. Figure 2.2 illustrates some
of Zorec's results.

Even more recently Vanbeveren (1977) analyzed the modification
that the equipotential surfaces undergo when taking into account not
only the radiation pressure from the two components of the system
but also heating effects from X-radiation, deviations from sychronous
rotation and a different rate of synchronization for the two components.
The deviations found from the conventional equipotential are large.
Observationally, HD 47129 provides a clear example of the effect of
the radiation pressure upon the gaseous structure in the system - actually, upon the direction of the gaseous streams and upon the density
distribution in the outer envelope (Struve, Sahade and Huang, 1958).

The conclusion is that any discussion that bases the arguments on
the shape and size of the zero-velocity surfaces should be done with
the uttermost caution keeping in mind the limitations involved. This
is particularly true in the case of systems with early type or X-ray
components as it is also true when one tries to determine masses on
the assumption that one of the stars fills exactly the corresponding
lobe of the inner contact surface, derived in the context of the restricted three-body problem. Sahade (1963) and Sahade and Ringuelet
(1970) were able to show that such a procedure is inadequate and could
give masses that are wrong by as much as 90% or mass-ratios that could
be up to 40% off if the actual radius of the star differs by only 10%
from the adopted size of the equipotential lobe.

If used properly, the equipotential surfaces can be and have been
extremely important for interpreting observations of close binary systems and understanding them in terms of evolution. In some cases
(Wood, 1946) they have been also useful for determining limiting
values of the mass-ratio when adequate spectrographic material is not
available. They should, however, be used with a full understanding
of the limitations set by the assumptions made when computing them.

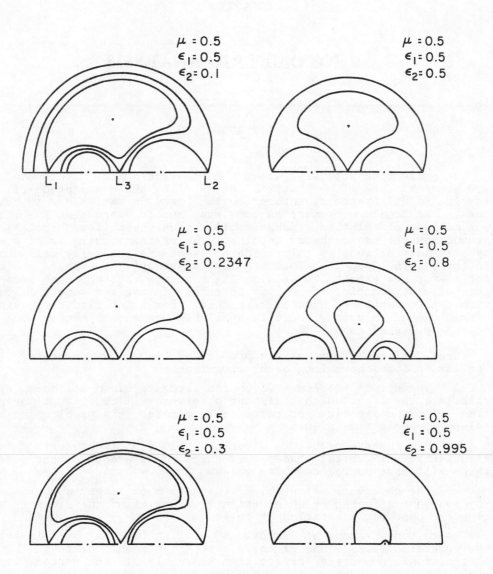

Fig. 2.2. Roche equipotential surfaces when the effect of
the radiation pressure from the two components are considered.
Zorec's nomenclature is different than the one used in the
Chapter: Zorec's ε is equivalent to the radiation pressure
parameter δ; Zorec's L_i's follow Pollard's (1966).

CHAPTER 3

PHOTOMETRIC ANALYSIS

1. INTRODUCTION

The present chapter will attempt an account of the development and present state of photometric theory; i.e., the description of any individual system by interpretation based on variations in brightness. The treatment cannot be complete, and in discussing the various methods of solution, emphasis will be on the philosophical background and the basic theory involved, rather than trying to give step-by-step instructions as to procedure; these are generally well described in the original papers. This summary is primarily for those who are not themselves experts in solving light curves, but who wish some general understanding of methods currently in use and some familiarity with the concepts and nomenclature used. For a detailed discussion of methods of solution, there is available the authoritative work edited by Tsesevich (1973).

Three different forms of the problem of light changes caused by stellar eclipses have long been recognized.

a. The direct problem. Given the elements of an eclipsing system calculate the light which a distant observer will receive at any given time. For the spherical or rectifiable models, this problem may be solved directly and uniquely.

b. The inverse problem. Given the values of light for a sufficient number of orbital phases, determine the elements of the model which will best reproduce these values.

c. The observational problem. Given a much larger number of observations affected by observational errors, find the elements which give the best representation of these.

The various methods of "solutions" of light curves now fall into two general classes. One of these, practicable only since the advent of electronic computers, treats simultaneously all the various factors causing brightness changes. The other much older methodology attempts to remove from the light curve all changes other than those caused by the mutual eclipses, and thus to derive a curve suitable for solution for the various elements mentioned earlier. This process is called rectification and the new curve is called the rectified light curve.

We shall start with a brief description of the original Russell method as applied to total-annular eclipses (Russell, 1912a). There are two reasons for this.

The lesser is historical. For many years, the published solutions were based on this method or some modification of it; for their proper evaluation, it is important to have in particular an idea of the assumptions involved. More important, its adaptation both to computer techniques, and in the form of the nomographs as developed by Merrill (1953b), led to methods in use today. Further, as pointed out by Koch (1970), in various later treatments, "the continuity of the terminology and the survival of many of the initially chosen parameters from 1912 to the present show clearly that all these efforts are really working in the lineage of Russell's conception of close binary models".

2.   THE RUSSELL METHOD

Russell's 1912 paper may be summarized as follows. First it was noted that the light elements - the period and some initial epoch - could be determined independently and as a consequence need not be included in the method of solution. In this paper, only well separated systems were considered so that interaction effects of all kinds were negligible and thus could be ignored.

Since absolute dimensions and luminosities cannot be determined from brightness changes alone, the combined light of the two components was taken as unity (i.e., $L_1 + L_2 = 1$) and the separation of the centers of the two components as the unit of distance. The sizes of the components, r_1 and r_2, are then expressed as fractions of their separation. The ratio of the radii is designated as k. Thus $k = r_2/r_1$, with r_2 being the radius of the smaller star; therefore, $k < 1$. In the very early literature there was a tendency to define k as the ratio of radii of the fainter to the brighter star; hence there may be found occasionally a published value of k greater than one.

The orbit of each component about the center of mass was assumed to be circular. In the case of a total eclipse L_1 and L_2 are determined before consideration of the other elements. For a final simplification it was assumed that the disks were uniformly bright, i.e., that the coefficient of limb darkening was zero. Later the case with complete limb darkening was considered, and much later Merrill, and independently Tsesevich, prepared more extensive tables for full and for specified intermediate degrees of darkening.

With the above simplifications, the problem was reduced to consideration of the eclipse of a uniformly bright circular disk by another circular disk. First considered was the case of a total eclipse. Even in this case, the problem of computing light changes as a function of phase is not an easy one. However, the real question was the inverse and more difficult one: given the observed light changes, compute numerical values for the elements which would produce these changes. The elements to be determined were:

$$r_1 = \text{Radius of the larger star,}$$
$$k = \text{Ratio of the radii} = r_2/r_1,$$

L_1 = Luminosity of the larger star,
(determined in this case directly
from depth of primary minimum),

i = Inclination or angle between the
perpendicular to the orbital plane
and the line of sight.

Taking θ as the longitude of the smaller star in its orbit as
measured from inferior conjuction - i.e., in practice, from the middle
of the primary eclipse - and δ as the apparent separation of centers,
simple geometry gives

$$\delta^2 = \cos^2 i + \sin^2 i \sin^2 \theta. \tag{1}$$

Next α is defined as the fractional area of the disk of the smaller
star eclipsed at a given phase. This can be read directly from the
light curve when the latter is expressed in units of intensity. Some
arithmetical manipulation and the introduction of an auxiliary function, Ψ, finally produced the expression,

$$\sin^2 \theta_1 = A + B \ \Psi \ (k, \ \alpha_1). \tag{2}$$

In this, A and B are constants which can be computed from the observed
phases at two arbitrarily chosen values of α. Ψ is a function only of
k and α_1 once these two arbitrary values are selected, and hence can
be tabulated. This if k is known, one can take tabular values of α_1
and use eq. (2) to compute the corresponding values of θ. In this
way, the light curve, during the eclipse of the smaller star can
easily be computed.

The trouble is, of course, that the value if k is not initially
known; this is one of the parameters to be determined. However, by
reading from the observed light curve a value of θ for a tabulated
value of α, k can be computed. By repeating this for a number of α's
and taking a suitably weighted mean for the various values of k thus
found, a light curve is computed which may be tested against the observations.

The curve thus determined has one disadvantage. By necessity it
must pass through the two arbitrarily selected points of the light
curve, thus giving these undue weight. In proper practice, however,
this is not the final case. Once an initial fit has been determined,
improvements are found by adjusting the parameters A and B in order
to get the best fit to the entire light curve. The final solution
may not pass precisely through either of the two initially chosen
points. The shoulders where the eclipse is beginning or ending and
the points just before and after totality are usually extremely sensitive to such adjustments, but no part of the curve is unduly favored.
This is emphasized here because the criticism of the dependence on
three points occasionally has been made by authors who failed to
understand properly the methodology involved.

With k determined, eq. (2) can be used to compute the phases of
the beginning of eclipse and beginning of totality (α = 0 and 1, respectively) more accurately than can be determined from the light
curve. The first phase corresponds to $\delta = r_1 + r_2$, and the second to
$\delta = r_1 - r_2$. Substituting in (1) and some trigonometric manipulation
leads to

$$r_1^2 \cosec^2 i = \frac{B}{\phi(k)} \qquad (3)$$

$$\cot^2 i = \frac{B}{\phi_2(k)} - A \quad , \qquad (4)$$

where $\phi_1(k)$ and $\phi_2(k)$ are tabulated functions. Solution of equations
(3) and (4) gives r_1 and i. The fractional brightness at totality
has given L_2. The desired parameters have been determined, and the
light curve is "solved".

We emphasize again that few, if any, computers would use this method
today for determining the elements. It is presented here to introduce
the concepts and terminology which are in use and as a basis for the
various modern methods which can be described only briefly. It might,
however, be an interesting "exercise for the student" to compare the
results using this method with those found using modern techniques.
A complete description of the development and use of this method in
more modern form has been given by Russell and Merrill (1952).

Russell also presented a means of solving partial eclipses. He
concluded that to the degree of accuracy of the existing observations,
the solution was indeterminate if only the light curve in primary minimum were available. However, if the depth of secondary eclipse has
been observed, another relation between α_o (the maximum obscuration)
and k could be written which made a solution possible, using another
set of tabulated functions (χ functions) developed for use with partial
eclipses.

This initial treatment of a greatly simplified situation was followed by a series of papers in which were considered the effects on
the light curve of orbital eccentricity, of the ellipticity of the
stars, of the radiation effect, and of limb darkening. Many years
were to pass before significant improvements were made.

3.   RECTIFICATION PROCEDURES

3.1 As time passed, of course, various modifications were suggested
which generally differed only in details of procedure. However, the
number of light curves showing unusual effects not accounted for in
existing models plus the general development of the theory of stellar
structure and atmospheres, eventually called into question some of
the fundamental geometrical and physical assumptions pertaining to
these models. Since the basic geometry had not changed, it was the
interaction (or proximity) effects connected with the rectification
which were first reconsidered.

The need for rectification had been early recognized observationally
and early discussed, (Pickering, 1880; Myers, 1898; Roberts, 1903;
Dugan, 1908, 1909, 1911; Stebbins, 1911; Russell, 1912b). Essentially,
observed changes in the light between eclipses can be used to compute
"adjustments" and a "rectified" light curve.

The two effects, which are usually the major ones, and which have
been most discussed and are probably best understood are known as

"ellipticity" and "reflection". The stars were first assumed to be
ellipsoids of similar shapes. In Russell's treatment, this rectified
curve was that which would be given by two stars, either uniformly
bright or completely darkened at the limb which showed circular disks
whose radii equaled the semi-major axes of the real stars and whose
surface brightnesses were those of the irradiated (and also the
eclipsed) sides of the true components.

The changes in the older rectification procedures concentrated
initially on the "reflection" or re-radiation effect and on the
presence of terms not predicted in the simple theory. Since rectification is usually treated by representing the portions of the light
curve outside eclipse by a truncated Fourier series, we find effects
which to some degree of accuracy will be represented over these intervals by sine terms and cosine terms. Since there is at present no
clear explanation for the sine terms, precisely how valid it is to
extrapolate these into the eclipse portion of the curves is an extremely troublesome problem for which there is as yet no clear answer.
Other phenomena, such as the gravity effect, the possibility of excess
light from an invisible companion, or the presence of shells and
streams were also considered. Along with these refinements came the
disturbing realization that, in many cases at least, the light curve
did not repeat precisely from season to season or even from night to
night, suggesting that overly-refined models might be largely ir-
relevant when applied to the acutal systems.

Discussion of rectification culminated after a number of developments in the treatment given by Russell and Merrill (1952). This was
based on earlier work by Eddington and Milne in their treatment of
reflection, and was slightly modified later by Merrill (1970). Especially in the case of extremely close systems (W Ursae Majoris type),
Merrill was able to show that serious errors could be introduced by
neglect of the higher harmonics and therefore could result in significant errors in some of the derived elements. Inclusion of terms in
both $\sin \theta$ and $\cos \theta$, as far as the fourth harmonic, were shown to
be necessary for a valid harmonic analysis of the regions of the
light curve outside eclipses. Russell and Merrill described in detail
an efficient method of rectification by graphical means, and have
thoroughly discussed the relevant factors including the gravity effect.
This work is recommended to those interested in further details.
Other authors (e.g. Budding 1973, 1974) discussed the use of high
speed digital computers in calculations of proximity effects.

In recent years, the subject of rectification has received more
attention from the consideration of the physical conditions in the
stellar atmospheres. Various authors (e.g. Russell, 1948a; Jurkevich,
1970) have considered the astrophysical significance of rectification.

It should be noted of course, that the complexity of the true situation is enormous. "Reflection" is more than merely absorption and
re-radiation and the spectral distribution of the re-radiated light
need not be that of the absorbed one. Rucinski (1970b), for example,
in addition to the thermal absorption from various sources, took into
account both scattering by free electrons (Thompson scattering) and
by neutral hydrogen atoms (Raleigh scattering) in his discussion of
photometric proximity effects in stars of early spectral type. The
true shapes of the stars will probably be neither ellipsoids or precisely that of the equipotential surfaces which are computed under
simplifying assumptions (as, for example, the neglect of radiation
pressure). Turbulence, meridional circulation, prominence activity,

the presence of circumstellar material, all make the real situation still more complicated. We will attempt to present what have been the most useful pragmatic approaches and to indicate some of the present types of study.

3.2. At least as early as Laplace, astronomers and mathematicians have considered the shapes of fluid objects under the influence of their own gravity, their rotation, and the tidal effects caused by the gravitational attraction of another mass. In his original papers, Russell drew on earlier work, especially by Myers and by Darwin. This approach in more modern form is found in Princeton Contribution No. 26.

There are two general effects on the light curve caused by the disturbed shapes of the stars as they move in their orbits around the common center of mass. One is the changing surface area from which the emitted light can reach the observer. In one of a series of papers on distorted polytropes, Chandrasekhar (1933) has shown that up to and including terms in r^3, the equilibrium configurations in close binary systems are ellipsoids and further that, except for rather extreme mass-ratios the ellipsoids are very nearly similar. Near the eclipses we see only the smaller ends of the ellipsoids (or other configurations) while midway between eclipses we receive light from a larger area.

A second factor is the gravity darkening; the regions of the stellar surface which are farthest from the core - i.e., near the ends of the longer axes - emit less radiation per unit area than other areas of the photosphere. It has long been realized (e.g., Takeda, 1934) that the total energy emitted per unit area should be proportional to the local gravity, and hence the term, "gravity darkening". (We should note that in the literature, the expression "gravity brightening" has also been used to describe precisely the same effect. Presumably, this is because at time of eclipse the gravity effect makes the limb of the star brighter than the center.) The effect of gravity darkening can double the observed light variation and requires a distinction between the "photometric" and the "geometric" ellipticity. We emphasize that the proportionality between gravity darkening and local gravity applies only to bolometric radiation and depends on the assumption of no internal circulation of matter.

We also note that in most cases the two components will not be of the same shape. This is treated in different manners in the various methods of solution. In using the Russell model, it has been shown that the light curve can be represented reasonably closely by two similar ellipsoids each having an ellipticity which is the weighted mean of the two actual stars.

3.3. One of the interaction effects in a close binary system is that caused by the radiation from each star which falls on the facing side of its companion. This phenomenon, which for historical reasons is termed the reflection effect, is extremely complex and its proper treatment is possibly the most difficult problem in the treatment of light curves. Fortunately, in many systems the effect is quite small, so that errors in its determination have only a minor effect on the derived elements. However, the other systems, if the observations are of high precision, the effects will be significant. In any case, the intellectual challenge presented by the reflection effect has caused many astronomers to attempt to understand it better.

   The study of the reflection effect for some time could assume only
specular reflection; even in this case, the mathematical treatment
was extremely complex and no truly rigorous solution was found. The
conventional theory, taking both reflection and ellipticity into
account, called for terms in cos θ and cos 2θ, where θ was the phase
angle from primary minimum measured as fraction of the period and
expressed in degrees. The cos θ term, as observed, was due to the
difference of the reflection effects of the two components; the cos
2θ was more complex; it was predominently due to ellipticity, but
included contributions from reflection. The assumption of sphericity
of the components for reflection, and similar prolate spheroids for
ellipticity, essentially assumed that higher harmonics could be regarded as negligible. It was realized that the assumptions involved
could limit the accuracy of representation especially in the case of
very close components. However it was soon obvious that the real
situation was even more complex; sin θ and even sin 2θ terms, for
which there was no simple theoretical explanation, appeared with
increasing frequency in the harmonic analyses as the observations
became more precise. A description of some of the difficulties en-
countered has been given by Koch (1970).

   The first really serious considerations were those of Eddington
(1926) and of Milne (1926) who assumed a parallel beam of incident
light to compute the phase variations. Further work was done by
Takeda (1934) and this was modified by Sen (1948) who assumed an incident beam which diverged from a point source. Russell (1948a) and
Russell and Merrill (1952) have produced further developments; the
latter gave a thorough discussion of the state of our knowledge in
1952.

   Another early contributor to the field was Hosokawa (1957, 1959,
1968) who published monochromatic analyses for a number of systems
and compared them with observations made at the same effective wavelength. Kopal (1959) attempted to allow in an approximate way for
the radius of the illuminating source, but Napier (1968) has given
reasons for doubting whether Kopal's method is suitable. A point
illuminating source was used by Sobieski (1965a, 1965b); he performed
monochromatic calculations and compared his results with a number
of observed systems.

   More recently, Napier (1968) discussed the photometric consequences
of the reflection effect and published tables for computation of the
incident fluxes. These are so extensive that, at least in the opinion
of one worker in the field "we can consider the geometric problem to
be solved (at least for not very close binaries)" (Rucinski 1970b).

   Other authors have considered proximity effects in other ways.
As an example, Chen and Rhein (1971) assumed spherical black bodies
with limb darkening. The temperatures of the facing surfaces were
then calculated as functions of the radii, the intrinsic temperatures,
and the limb darkening of the two stars. The light curves (neglecting
eclipse effects) were then calculated as functions of wavelength and
orbital inclination by the use of numerical integration. Direct computation from this simple model allows comparisons with some observed
binary systems. This paper used the results of a preceding one (Chen
and Rhein 1969), which described the method of calculating the distribution of temperatures on the surfaces of the components. Using
this method, monochromatic reflection effects could be computed.
The calculations were made as a function of phase for selected sets

of eight parameters; namely, the radii, temperatures, and limb dark-
enings of the two components, the inclination of the orbital plane,
and the wavelength of the observations. The light variation included
radiation from the penumbra as well as from regions full illumination.

Other authors who have made useful contributions include Pustylnik
(1967), Ureche (1972), and Ovenden (1970); the latter has discussed
the effects of hot spots and circulating currents in the atmosphere
of the reflecting star.

Rucinski (1969a, 1969b, 1970a, 1970b, 1971, 1973) has published
a series of papers considering proximity effects. His results are
worth noting. In the 1969b paper, for example, he found that the
efficient subphotospheric convection which extends throughout most
of the envelope of the late type star, would tend to decrease the
effective bolometric albedo to approximately half the purely radiative value. The theoretical value of the albedo depends on the mix-
ing-length/scale-height parameter, and on the value of the temperature
excess at the point where convection sets in. This idea of sharply differing albedos for early and late type systems was supported by Naiper
(1971) who repeated some earlier statistical investigations using "more
precise" theory. Contrary to some earlier work, the results show
that the albedos of early-type binaries agree broadly with theory,
having a mean albedo of 0.99. However, those of the later-type are
considerably less than unity, thus confirming the prediction made
by Rucinski. According to this, the cooler secondary components have
the structure of their deep convective envelopes changed by the incident radiation so as to depress the radiation from the interior; for
radiative secondaries without deep convective zones, this is not the
case.

If there is indeed a relation between the value of the albedo and
the mixing length parameter, there is the possibility of an independent determination of this parameter for subgiant secondaries. Thus
there may be an opportunity of using the reflection effect to gain
further knowledge of the structure of late type stars.

It would, however, be premature to state that the question is
entirely settled at this time. Pustylnik and Toomasson (1973) point
out that departures from LTE in the atmospheres of the cooler components of Algol-type binaries are probably very large indeed. The
incident radiation of the primary star is comparable to the normal
flux from the secondary. They conclude that temperature inversion
and an abnormal law of limb darkening will be the most serious consequences if the ratio of the effective temperatures exceeds 2.5. A
great deal of active work is going on in this extremely difficult
area.

4.   CURRENT METHODS OF SOLUTION OF LIGHT CURVES

4.1. Two general philosophical approaches to methods of solution
have developed. The first treats separately the non-eclipse factors
effecting the light curve and then computes elements of the system
from the rectified curve. Its further development has again been
separated into two approaches. In its graphical form it has culminated in the nomographs (Merrill, 1953b). It has been adapted to
machine treatment by various authors - e.g., Proctor and Linnell (1972)

or Budding (1974) who used a variant of it developed by Kopal.

The other approach is the development of a non-spherical model by what has become known as the "synthesis" method, first suggested by D. B. Wood (1962). This latter approach attempts to determine simultaneously all pertinent parameters. It has become practicable only since the development of electronic digital computers.

We shall first discuss in condensed form the use of nomographs as described by Russell and Merrill (1952). We shall then describe one method using the Russell model and finally two of the most generally used of those which have abandoned this approach.

4.2. The nomographs present a graphical method which is extremely valuable for a preliminary solution. It has much to recommend it as an initial approach. Many of the other methods require preliminary elements and the closer these are to the final values, the more rapidly the "final" solution can be made. In other cases the observations may not be numerous or precise enough to warrant refined computation, but may still permit the general nature of the system to be found.

The tables and methodology of the nomographic solutions have been derived for the spherical model using essentially the same assumptions as in Russell's original paper. The sole addition is that of limb darkening treated with the usual cosine law.

Normally a photometric solution requires a series of approximations appropriate to the individual case. The nomographs were constructed to facilitate both the initial and the later stages of the approximations and, with the simple tables appended to them, to provide for study of any desired choices of limb darkening on either component. They can, of course, in no way remove discrepancies caused by the departure of the actual system from the assumptions inherent in the rectifiable model.

A preliminary solution, which may be made with great rapidity will tell whether each eclipse is partial, total, or annular and gives parameters from which the elements of the system can be computed to the accuracy permitted by the model and by the extent and accuracy of the data used. Further, the nomographs permit rapid determination of the changes in the elements produced by permissible variations in the input data, as for example, in the depths of the minima which are not determined with infinite accuracy. When the solution is indeterminate, or practically indeterminate, this is clearly shown and the range of uncertainty indicated. Finally, there are cases in which no solution is possible; this is immediately demonstrated.

The first step, of course, is to plot the light curve and to study it carefully for effects not covered by any normal theory - e.g., "humps" found on the light curves of stars of the RS Canum Venaticorum group. The light curve should then be rectified by methods described in Princeton Contribution No. 26 or by other appropriate means. The rectified light curve should then show only changes caused by the mutual eclipses, and should include indications of maximum and minimum permitted ranges of the eclipses and a mean or "best" set of data for entry into the nomographs.

The use of the nomographs involves the use of the χ functions, which are determined by the shapes of the minima and of a relation depending on the depths of the minima (in the case of a circular orbit). The χ functions were originally tabulated for solution of

partial eclipses, but they were extended by Merrill's tables for
total-annular solutions as well. Derivation and discussion of these
functions have been given in Princeton Contribution No. 26.

To use the nomographs, we find directly from the light curve two
parameters computed from the rectified depths of the minima and
another which depends on its phase at two specified fractional depths
i.e., a "depth" relation and a "shape" relation. The shape relation
dictates the selection of a χ curve of some specific value from among
those plotted on the nomograph. The depth relation defines a straight
line. These two lines will usually intersect at one, (or possibly
two) places; the value of k (and of p) can then be read directly from
the nomograph. The coordinates used in preparing the nomographs were
chosen so that by reading their values at the point of intersection,
the values of α_0, L_1 and L_2 are easily computed. The requirement
$L_1 + L_2 = 1$, give a check against errors. The remaining elements are
computed from the relations given earlier. Note that this is a preliminary solution. The χ functions depend on selected points on the
light curve. The test is always the fit of the computed curve to
the entire run of observations.

Russell and Merrill have discussed various types of indeterminancy
sometimes shown by the nomographic solutions. These are not the
results of this particular technique. Frequently, they are in the
nature of the systems themselves and no method of solution can remove
them. Note further that if the true system departs widely from the
rectifiable model, the nomographs may yield unreliable results, but
in this case, the departures must be very great indeed. One virtue
of the nomographic approach is that when indeterminancy is truly
inherent in the system, this is clearly shown.

One type of indeterminancy, in particular, merits discussion.
This arises when no intersection occurs; the depth line passes below
the lowest point of the shape curve. One treatment which has been
used here is to note that if there is some source of "third-light" -
an invisible companion, faint nebulosity, a shell or ring - the observed depths would not be the true eclipse depths. By adding the
proper amount of "third-light" the depth curve could be raised until
it was tangent to the shape relation and a limiting solution made.

When this was first suggested for one specific case, it did not
seem unreasonable. However, the number of cases where added light has
become necessary makes this treatment increasingly dubious. The
amount needed to force a solution is frequently large (up to 1/3 of
the combined light of the system), yet it is not observed spectrographically or in usual color effects. For one individual system
reasons to rationalize this could no doubt be found, but such cases
should be extremely rare. Further, we cannot be sure that the amount
of extra light present is exactly that needed to raise the depth curve
to tangency. If more is present, a different solution will be found.
There is an even more disquieting thought. If a significant amount
of third light is indeed present in many systems, then it presumably
is also present in many for which a "solution" has been found, and
the elements thus determined are not the correct ones. It seem safe
to conclude, in the absence of spectrographic or other evidence, that
a considerable amount of extra light is not present in most cases, and
that when a solution is not possible without it, this is because the
real system does not conform in this case to the assumed model.

The next question is the proper treatment following the nomographic approach. In some cases, especially when the observations are scanty, of low accuracy, or poorly distributed, it will be clear that no amount of more complicated treatment can produce more reliable results and, until a better light curve is produced, we will have only a "low grade" solution, which, however, can frequently be useful in statistical studies, planning observing programs, etc. as long as it is realized that it is low grade.

In other cases, the nomographic approach may indicate (though lack of a well defined intersection) that the nature of the system and its orientation to us is such that no determinate solution is possible by this or any other method. An obvious case is that of two shallow partial eclipses when no spectrophotometric data are available.

For systems in which the preliminary solution indicated that more rigorous treatment was desirable, the most obvious approach in the 1950's was an intermediate solution utilizing either the ψ or the χ functions on normals of a few observations each, followed, if warranted, by a least squares solution utilizing any of the several approaches and sets of auxiliary tables available.

Today, one of the various methods using modern electronic computers can be applied. Precisely which one is a difficult question. The British poet R. Kipling once wrote "There are nine and forty ways of constructing tribal lays, and every single one of them is right". There are not - yet, anyway - quite that number of methods of solving light curves, but many ingeneous approaches exist. Indeed which is "right" may well depend on the nature of the system discussed or on the type of computer and amount of time available. The following discussion is intended to be illustrative rather than exhaustive.

4.3. The observational problem (obtaining the parameters from the observations) can be handled by the application of the method of differential corrections, using the expression for the total differential of the light values as the equation of condition for a linear least-squares analysis. This requires a good "preliminary" solution.

One modern example of its utility is shown by a group of computer programs developed by Proctor and Linnell (1972). This rectifies individual observations according to the Russell model and produces by differential corrections a "final" set of elements. Among other advantages, the following are cited (1) the same set of equations apply to partial as well as to complete eclipses; (2) the coefficients of the differential corrections in the normal equations are independent of observational error; and (3) the effects of orbital eccentricity can be included directly.

It should be made clear that no computer methodology is a substitute for judgment and experience and none can be used blindly. If two parameters are very highly correlated, a direct solution into which both enter cannot be found. Various writers have emphasized that, for the final elements to be accurately determined by least-squares adjustment, the initial parameters must be within a few standard deviations of the final values.

4.4. The limitations of the basic rectifiable model theory in handling in detail refined observations are now generally recognized. Indeed, it is rather surprising that a model suggested in 1912 lasted so long and is still so useful when properly applied. As Russell and Merrill have pointed out, in a typical Algol system the small, less-massive component should be nearly spherical while the other is more elongated. The "ellipticity effect" is now known to be much more complex than originally assumed; we have already mentioned some of the difficulties involved in treatment for "reflection". Actually, the Russell model could be modified to allow for difference in shape of the components and an approach has been made by Horak (1968) with his sphere and ellipsoid model. Some astronomers feel, however, that for this modification to be meaningful we should understand the gravity effects better than we do at present.

Critics of the rectification procedure argue that harmonic analysis and rectification separate the treatment into two parts in an artificial way. If troubles were found in the solution, it was much too laborious to repeat, especially with methods available in earlier days, alterations in the rectification. They also object to the lack of meaningful error estimates for each individual parameter, and urge the simultaneous treatment of all. In any method, of course, realistic error estimates will be difficult when dealing with so many interdependent parameters.

Of the various methods developed, we will summarize those of Wilson and Devinney (1971) and of D. B. Wood (1972). These are now used by many observers discussing their own observations, and each method has special merits to commend it.

4.5. Wilson and Devinney presented in 1971 the first general synthesis method of computing light curves of close binary systems and applied this to MR Cygni. Effects of rotational and tidal distortion, reflection, limb darkening and gravity darkening were all included. The "observational" problem (derivation of elements from the observations) was solved by differential corrections; probable errors were computed for all adjustable parameters. They noted that future refinements could easily be incorporated into such a flexible procedure. This model as first presented deserves consideration in some detail.

In the case of tidal distortion, the model is that in which the Roche equipotential surfaces are computed on the assumption of complete central condensation of the components, synchronous rotation, and a circular orbit. The selected parameters are (1) the inclination, (2) and (3) the monochromatic luminosities of each component, (4) and (5) the coefficients of limb darkening, (6) and (7) the gravity darkening exponents, (8) and (9) the temperatures, (10) the mass-ratio, and (11) and (12) the two surface potentials. Assuming equal gravity darkening for the two components reduces the number of adjustable parameters by one and, since in almost all well observed cases, the temperature of at least one component is known from the spectrum, the number of adjustable parameters is ten. If a spectrographic mass-ratio is available, they are reduced to nine. The number of adjustable parameters is not permanently fixed. Third light was an original parameter and two bolometric albedos were added in their paper on Algol. The results of the above application is a set of corrections to the starting elements, giving the most probable values for the elements and the probable errors of each. As the program is written, any subset of the elements can be held fixed while the others are adjusted. The use of the nomographic (or other suitable) technique to obtain the preliminary elements reduces the possibility of finding a local minimum which conceivably would not be the deepest minimum in the multiparameter space.

The determination of the light curve begins with computing locally on the distorted components all the pertinent physical quantities.

We thus have a large number of individual surface elements with a
relatively uniform distribution over the surfaces of the two components. By summing the flux from these visible to the observer, the
total observed flux is computed. Further details are given in the
original publication.

Note that the treatment can handle "over-contact" cases as well as
detached or semi-detached systems. In its latest version it treats
non-synchronous rotation, computes radial velocity as well as light
curves, treats rather thoroughly the reflection effect, and permits
the use of either stellar atmosphere or black body radiation physics.
It permits the simultaneous solution of light curves in many wavelengths, while constraining parameters which should not be dependent
on wavelength (e.g., mass-ratio, inclination) to have identical values
for all. It has a total of seven solution modes which makes it possible to use the same program for quite different types of binaries.
It merits serious study by any computer of light curves.

4.6. The second method we discuss here is that of D. B. Wood (1971).
As with other examples of the direct approach, this discards the
spherical model and the concept of rectification and attempts to take
into account in a more nearly exact manner the complexities of a
binary star system. The various perturbations are represented by
straightforward physical models. Thus one can build in a computer
a detailed model of an eclipsing binary system and study "with great
accuracy" effects of nonsphericity, gravity darkening, reflection,
limb darkening, and orbital eccentricity. The basic philosophy is
the same as that of Lucy, of Hill and Hutchings, and of Wilson and
Devinney.

In Wood's model, the parameters are divided into three classes -
orbital, geometric, and photometric. The photospheres of the two
stars are assumed to be triaxial ellipsoids (in general, not of the
same shape). The period of rotation equals the period of revolution.
The program can treat eccentric orbits.

In treatment of the reflection effect the geometrical complexities,
especially in the penumbral regions, are removed by a vector treatment of the problem (see Chen and Rhein, 1969). This was selected
in preference to a more sophisticated astrophysical treatment because
of the problem always facing a computer without unlimited machine
time available - the trade off or decision between accurate integration of a relatively simple model or poorer integration of a more
sophisticated model. In reality, of course, neither of the models
may represent the actual atmosphere.

The total observed luminosity of each component is that given by
an integral over the apparent disk of the sum of the emergent intensity and the reflected intensity, duly corrected for limb darkening
and gravity brightening. The luminosity of the system at any time
is the sum of these, less the light lost in eclipse, when the equation
is integrated over the overlapping area. In his paper, Wood compared
his computations to five observed light curves with favorable results.

Let us again point out that no method of solution can be used
blindly. One example of this is shown in D. B. Wood's (1976) analysis
of CD Tauri. Starting with assumed limb darkening coefficient of 0.6
he found a solution which fit the data satisfactorily, but which gave
darkening coefficients for each star of 0.26 in the yellow and 0.36
in the blue. These values are much smaller than expected from model
atmospheres. By starting with theoretical darkening values, a much

more satisfactory fit was obtained. Wood comments, "This is a good
example of the problem of local minima in the multi-parameter phase
space in which the computer solution is performed". Hence, there is
always a danger of a particular solution not being unique. Comparison
of the results of different models is highly desirable.

4.7. We should mention that there are various methods which depend
on the Fourier transform of the light curve. Early work was done by
Kitamura (1965) and by Mauder (1966), and a somewhat different approach is being developed by Kopal (1975). In this method frequency
is taken as an independent variable instead of time. Ideally, this
offers a chance of separating "noise" from other effects.

5. SUMMARY

In summary, the astronomer wishing to solve a light curve has the
choice of two basic philosophies. He can first remove the non-eclipse
effects from the curve and then proceed to solve the rectified light
curve, or he can select one of the methods which treats all the parameters simultaneously. In the first case, he has the choice between
graphical methods which have reached their highest development to
date in the nomographs, or he can follow the computer techniques as
presented by Proctor and Linnell, Budding, and others.

In most cases, we prefer initial use of the nomographs to obtain a
preliminary solution, to get an idea of the range of determinancy,
and to decide whether a more refined solution would be more meaningful. There is no need to fill the literature with solutions carried
out to the third decimal place, when the second is meaningless. A
well observed light curve is not always sufficient; some systems are
such that precise determination of elements from light changes alone
is not possible. The investigator should, in particular, be aware
of the basic assumptions underlying each method and the extent to
which they are likely to be violated by the real system in each particular situation.

It may be instructive to try more than one method to compare the
conclusions which different treatments may draw from the same observations; if methods involving rectification are used, it must be remembered that alternate values of rectification must also be considered.
The same precautions apply to those who compile tables of elements who
should take care to explain fully the procedures of selection.

CHAPTER 4

W URSAE MAJORIS SYSTEMS

We make a sharp distinction between W Ursae Majoris sytems and binary systems showing light curves of the W Ursae Majoris type. The latter designation refers to all systems showing minima of comparable depth and an appreciable amount of continuous light variation between minima. Here, we confine the discussion to systems with periods less than one day and spectral types of F or later. This necessarily excludes discussion of hotter contact systems on which stimulating work is being done by K-C Leung and others, but it is necessary if we are to discuss a homogeneous class.

The W Ursae Majoris systems have long been of special interest. Their space densities are possibly the highest of any type of binary stars. Their short periods (less than a day and usually less than 12 hours) have made them favorites of observers who wished to obtain a light curve with minimum effort. Spectral classes run from A8 (rarely) to K0 and the spectral types of the components tend to be similar. The continuous curvature of the light curve between eclipses clearly indicates large interaction effects. The eclipses tend to be of comparable depth; the depths range from a few tenths of a magnitude to a little more than a magnitude and there is some theoretical reason for believing no depths greater than about 1.25 magnitudes should exist (Lucy, 1968b). Asymmetries in light curves are common.

In general, the components are located on or near the main sequence; absolute magnitudes run from about +3 to +6. However the secondaries are generally slightly overluminous for their masses and the primaries underluminous in terms of the conventional mass luminosity relation, according to Kitamura (1959) and Osaki (1965). Combined masses run roughly between $0.8 \mathfrak{M}_\odot$ and $2.8 \mathfrak{M}_\odot$. A typical mass-ratio is 2:1, (cf. Struve, 1950a), although the luminosities are comparable; equal mass ratios are not found. In some cases there is evidence for mass ratios which are considerable larger than 2. A period-color relation seems to exist (Eggen, 1961, 1967), in the sense that the redder the system, the shorter the period. In common with other close double star systems, they are at least extremely rare in globular clusters and no certain case is known.

Their evolutionary status has presented certain problems. Perhaps the major historical puzzle was the association of nearly equal

34

luminosities of the components with significantly different masses. However, this feature presents no particular problem to present models.

Until about 1968, solutions were carried out by conventional methods based essentially on the Russell model, although the model was clearly not designed for such systems and some of the assumptions inherent in it (especially that of similar ellipsoids) were almost certainly violated. This was of particularly serious consequence in these systems because of the large size of the interaction effects. To summarize, it was generally found that the simple harmonic analysis approach embodied in that model and others was inadequate for dealing with these systems and elements derived by its use were unrealistic. Merrill (1970) was able to show that serious errors in the computed elements could occur unless both $\sin \theta$ and $\cos \theta$ terms, up to and including terms in $\sin 4\theta$ and $\cos 4\theta$ were included in the analysis. Further suspicion was cast by the changing shapes of the light curves which frequently varied from season to season and in some cases within a given observing season.

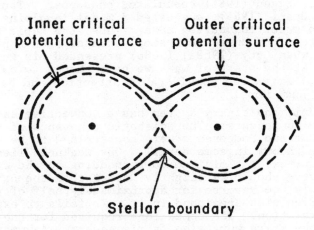

Fig. 4.1. Schematic of a W Ursae Majoris system.

In addition to the difficulties with rectification, another paradox defied solution. If the stars were indeed "contact", they would fill their critical lobes and this meant,

$$R_2/R_1 = (\mathfrak{M}_1/\mathfrak{M}_2)^{0.46} \tag{1}$$

where the R's are the radii and the $\mathfrak{M}$'s are the masses. Actually, there was not sound evidence that the stars did <u>precisely</u> fill these lobes, but the approximation was close enough for the purpose. The paradox came when the mass-radius relation was written,

$$R_1/R_2 = (\mathfrak{M}_1/\mathfrak{M}_2)^n \qquad\qquad (2)$$

For stars on the lower part of the main sequence, n is very close to unity. The only way both equations could be solved was for the masses to be equal, and this was absolutely forbidden by the radial velocity observations. Various efforts were made to explain the existence and nature of these systems.

For example, Huang (1958) in discussing the mechanism for energy transfer suggested corpuscular heating. Sahade (1960b) suggested mass transfer as did Kraft (1962), although Kraft (1967) later questioned this. One year later, Kitamura (1960) presented an alternative suggestion, namely that of contraction and mass accretion by the secondary component. Eggen (1961) speculated on a post T Tauri stage of contraction. Osaki (1965), suggested a model involving circulation currents in an outer convection zone. Thus, the topic was already being considered from an astrophysical, rather than a purely geometric, point of view when Lucy (1968a, 1968b) presented his intial model. Coming almost at the start of what we have chosen to call the modern era, this model attracted considerable attention and has come to be known as the "Lucy Model".

In this model, the binary system has a convective envelope which surrounds both components. The adiabatic constants of the convective envelopes of the two components were constrained to be equal. Lucy worked out his theory in some detail. One weakness lies in the predicted light curves; the minimum corresponding to the eclipse of the more massive star should be slightly deeper than the other. The situation is exactly the reverse for approximately half of the systems, and other difficulties are found in it. It fails to explain the period-color relation. The conditions required for the internal structure of the stars were also fairly severe. Asymmetries in the light curve were not explained nor was the reflection effect treated. In addition to the equal entropy restraint, both components were assumed to be at age zero. Clearly much work remained to be done, but this approach suggested a treatment which was astrophysically realistic. Lucy's model stimulated a great deal of work in the field, most of which can only be summarized briefly.

Moss and Whelan (1970 retained the common convective envelop, but used more recent values of the opacities. They found that zero age components were possible only for extreme population I compositions and, for at least some observed systems, they were completely unable to explain their location in the period-color diagram. Other theoretical studies were by: Hazelhurst (1970) who found consistency only when the primaries were highly evolved and the secondary of age zero, and who pointed out that TX Cancri (in the Praesepe cluster) was a complete enigma; van't Veer (1972) who tried to explain period changes in contact binaries as a result of mass flow; Mochnacki and Doughty

(1972a, 1972b); Rucinski (1974); Biermann and Thomas (1973) who relaxed the requirement of equal entropy, but found even bigger disagreement with the observed light curves; their method was further extended by Vilhu (1973a). We might note that Whelan et al. (1973) were able to construct an age-zero contact-binary model for TX Cancri.

Rucinski (1973) has developed two methods of determining the geometrical elements from the light curve. He first computed a grid of 48 theoretical contact model light curves for different combinations of these elements which he defined as the mass-ratio (q), the fill-out parameter, (f) and the inclination (i). The fill-out parameter is defined as $(C-C_2)/C_1-C_2)$, where C is the Jacobian constant for the common stellar envelope and C_1 and C_2 the Jacobian constants for the surfaces passing through the inner and outer Lagrangian points. While Rucinski could get solutions by "force-fitting" the theoretical curves to the observed ones, he was disturbed by the fact that the transit eclipses were not properly represented. Hence, he considered other methods. The Fourier coefficients of the cos 2θ and cos 4θ terms, A_2 and A_4, were used for determinations of f independently of q and i. This depended only on the ratio: A_4/A_2. Thus if the mass-ratio is known by some independent method (such as velocity curves) it is possible to compute the inclination. Rucinski also presented a method of determining the elements by making use of the amplitude and the half width of the eclipse in the circular orbit case.

Finally, two features of the W Ursae Majoris systems deserve special comment. One is the apparent division into two classes which are commonly called A and W. Originally the division was merely on whether the more or the less massive component was eclipsed at primary minimum, but other differences have now been noted.

In general, stars in systems of type A tend to have slightly higher luminosities and masses and to be of slightly earlier spectral types than do their counterparts in systems of type W. Phenomena such as asymmetric light curves or changes in light curves tend to be moderate or even absent entirely in type A systems while they are prominent in almost every W light curve. On the whole, type A systems have smaller mass ratios, although there is considerable overlap. The envelopes seem to have a higher degree of contact in the A systems, and their light curves conform much more satisfactorily to the contact model. The A systems, can be explained by energy exchange in the adiabatic regions of the common convective envelope while in the W systems, superadiabatic regions are postulated. The orbital periods of the W systems are continually changing, while the A group has systems with sporadically changing periods. That is, the peculiar systems seem to be concentrated rather heavily in the W group.

. All of this has suggested to some authors that many if not all of the A type systems are evolving on the nuclear time scale with a structure similar to Lucy's model, and that if we define the mass ratio as $\mathfrak{M}_2/\mathfrak{M}_1$, stable systems can form and exist only below a certain critical value.

In a later paper, Rucinski (1974) considered the differences in depth of minima and in temperature deviation (from Lucy's original model) for the secondary star and included the reflection effect. He introduced a new parameter, X, which depends on the temperature deviations of the secondary components from those of the conventional contact model. This clearly separates the A and W systems and supplements the older division which depended on the relative depths of the transit

and occultation eclipses; only a few systems are re-classified but
the distinction appears to be a more fundamental one. A relatively
crude analysis of the period changes suggest that significant exchange
of mass in the W-type systems occur in the Kelvin time scale (roughly
10^7 years) while in the A-type systems, the time scale is at least
an order of magnitude greater, in agreement with the evolutionary
discussion in the preceding paper.

Rucinski proposes that the internal structure of an A type system
consists of a shallow convective zone surrounding a radiative envelope common to both stars; a W-type system would have a deep convective envelope quite similar to that originally proposed by Lucy.
This is in contradiction to models assuming unequal entrophies for
the components.

The entire picture is, of course, an extremely complex one. For
example, various authors have constructed models that evolve to
"marginal contact", then break contact, and evolve back to contact
again. To check observationally whether these models correspond to
reality may not be an easy task.

In a recent paper, Shu et al. (1976) reviewed the problem of the
interior structure of contact binaries and suggest a solution to some
of the difficulties. They suggest a contact discontinuity between
the lower surface of the common envelope and the critical zero-velocity lobe of the star of lower surface temperature. Horizontal (i.e.,
perpendicular to the local direction of the effective gravity) exchange of mass and heat occurs to any extent only in that part of the
envelope above the lower critical surface. Here the equipotential
surfaces are common to both stars and the gas can easily move hori-
zontally. Inside their respective lobes, the stars are "nearly de-
coupled" and this suggests a nearly discontinuous behavior across
the inner surface. This model does not require a basic distinction
between a radiative or a convective envelope and it does not involve
special combination of modes of energy generation.

The age and the final evolutionary states of W Ursae Majoris stars
is at this writing far from firmly settled. A related question is
the existence of massive contact systems. Leung (and others) have
applied the contact model to such systems having hot stars with radiative envelopes.

We should end with a note of caution. We have summarized as well
as we could in any reasonable space some of the recent developments
in the field. Many excellent papers have been omitted. While an
enormous amount of work has been done recently, usually adapting
some version of the Lucy model, it is by no means certain that this
will be "the" model of the future. The hard test of detailed comparison with precise observation still presents many problems.

Many years ago, Struve (1950) noted that if both components were
really in contact, the wings of the absorption lines would touch at
quadratures and that this was not true for W Ursae Majoris. Later,
Koch (1968) pointed out that photometric analysis for "the most determinate" cases indicated semidetached binaries and preliminary results
using narrow-band filters gave light curves which were "significantly
different" from those made with broad-band filters. In a more general
discussion, Koch (1972) discussed anomalies in the photometric behavior of close binaries and urged caution in their interpretation.

We emphasize that to be fully accepted, a model not only must conform to the laws of physics, but must predict results which agree with the observations to within the accuracy to which the latter can be trusted. The volume of work now going on indicates that this is not yet the case.

CHAPTER 5

OBSERVATIONAL EVIDENCE FOR
GASEOUS STRUCTURE

1. INTRODUCTION

When discussing the third epoch in the development of the field of
close binaries in our introductory Chapter we referred to the Struve
revolution. As a result of it we now picture (Batten, 1970, 1973a, 1973b
Sahade, 1973) an interacting binary as a physical system of two stars
with a gaseous structure where we recognize the following elements,
namely,

a gaseous stream, normally flowing from the less massive component
towards the companion,

a circumstellar envelope, around one of them, normally around the
more massive component, and

an outer envelope which surrounds the entire system.
The circumstellar envelope could have either a rather flat or spherical distribution around the star. In the former case we would talk
about gaseous rings or about disks, depending upon whether the density
is small or large. We may even have a circumstellar envelope with the
matter concentrating on a disk and thinning out in all directions and
having overall a spherical distribution.

The phenomena connected with gaseous rings and disks are largely
found in spectroscopic binaries which are at the same time eclipsing
stars, a fact that suggests that the matter in the rings and disks is
primarily concentrated in the orbital plane and has essentially a
flat distribution. Evidence for gaseous formations that are spherically distributed are found in objects which, of course, do not need
to undergo eclipses.

There are already several reviews that discuss different aspects of
the problem of the gaseous structure in close binaries. As early as
in 1955 Wood (1957a) dealt with the photometric evidence of instabilities in eclipsing binaries, in 1968 Sahade (1969) summarized the observational spectroscopic evidence for mass loss in close binaries and,
more recently, Batten (1970, 1973a, 1973b) has reviewed our knowledge
regarding the circumstellar matter in close pairs, while Huang (1973)
undertook an overall examination of the problems of the envelope in
eclipsing binary systems. Although there is no need to repeat again
everything that has been said, it might be useful to condense in this

Chapter all the available observational evidence for the existence
of the mentioned three elements in the gaseous structure of interacting binaries.

2.   EVIDENCE FOR GASEOUS STREAMS

2.1.   One piece of evidence comes from <u>extra absorption lines</u> that
appear in the spectrum at certain phases, superimpose upon the normal
stellar absorptions and produce sometimes broad lines that yield radial
velocities which do not reflect purely orbital motion and give a <u>dis-
torted velocity curve</u> for the corresponding component. This is so in
interacting Algol-type binaries. The distorted velocity curves that
result are peaked immediately after about phase 0.75 P and give rise
to the Barr effect, already mentioned in the introductory Chapter.
The distortion shown by the velocity curve of U Cephei (Fig. 5.1) is
typical and Struve (1944a) suggested that it resulted from the existence of a gaseous stream similar to those also postulated by Struve
(1941) in β Lyrae to account for the so called "satellite" lines -
which are now interpreted in a different way (Batten and Sahade, 1973)
as we shall mention in Chapter 10 - and in SX Cassiopeiae (Struve,
1944b). It was further postulated that the "orbital" velocity curve

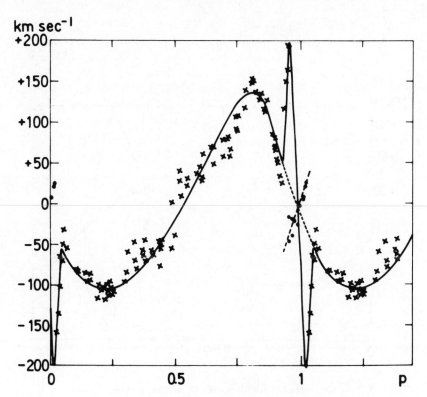

Fig. 5.1.   The velocity curve of U Cephei as determined by
Struve at the McDonald Observatory.

was actually sinusoidal, as was suggested by the light curve, and
that there were superimposed effects of the radial velocity of moving
masses of gas. Fig. 5.2 illustrates the velocity distribution found
by Struve in SX Cassiopeiae and his interpretation of the velocity
curve.

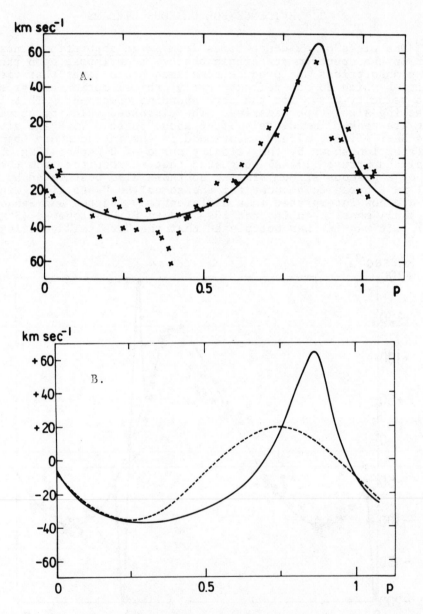

Fig. 5.2. The velocity curve of SX Cassiopeiae.
         a) The velocity distribution.
         b) Struve's interpretation of the velocity distribution:
             the orbital motion is represented by the dashed curve.

AU Monocerotis (Sahade and Cesco, 1945) was the first case where it
was found that the effect of the distortion was present in the H lines
and not in HeI, and the same was later found to be true in U Coronae
Borealis (Sahade and Struve, 1945), a fact that is related to the
physical condition of the gaseous stream.   In the phase interval from
about 0.75 P to about 1 P where the velocities derived from the H lines
are more positive than those derived from HeI, the H lines in AU Monocerotis are broader than in the rest of the cycle.   This suggests that
the stellar lines are blended with lines that correspond to higher
velocities of recession and originate in a mass of gas seen projected
upon the receding hemisphere of the primary star and moving towards
this star.   Struve (1949) estimated that the velocity of the stream,
in the case of U Sagittae, is of the order of 200 km sec^{-1}.

2.2.   If there is a gaseous stream in a system, one should expect to
find, at quadratures, evidence for matter - perhaps even a concentration of matter, if matter should flow through the area containing Lagrangian point I - between the two stars.   This is indeed the case in
β Persei (Algol) (Struve and Sahade, 1957; Sahade, 1958a; Andrews,
1967), in UX Monocerotis (Struve, 1947) and in U Cephei (Batten and
Laskarides, 1969), where H emission is detected at elongations, though
the intensity of such emission appears to be different at both quadratures.   Similar emission should be responsible for the weakening
at quadratures of the H stellar absorptions in SX Cassiopeiae (Struve,
1944b) and RX Cassiopeiae (Struve, 1944c); it must result from superimposed emission.   It is worth noting that this weakening is more
conpicuous at the quadrature that precedes primary minimum.   The
phenomenon is the same and must be related to the density of the
gaseous matter facing the observer, which must be larger in the phase
interval in which the star towards which the stream is directed has
velocities of recession.   We are dealing with the same phenomenon
that produces the kind of variations in the intensity of the absorption lines discovered by Bailey and found in several spectroscopic
binaries, as mentioned in the introductory Chapter.   However in this
case one must be dealing with continuous radiation from the stream
as it was postulated by Sahade (1959) in the case of UW Canis Majoris
to account for the disappearance of the lines of the secondary component during the phase interval where the radial velocities from
this component are of recession.   Continuous radiation from the
stream should account for the veiling of the spectra in HD 471 (Sahade,
1962) and in V453 Scorpii (Sahade and Frieboes-Conde, 1965) at the
phases where we should be seeing the gaseous streams projected on
the star towards which the stream is moving.

Table 5.1 summarizes the cases where such a variation in intensity
of the stellar absorption lines has been detected, with the name of
the observer and the year.

In short, H emission at quadrature, yielding the appropriate radial
velocities, and weakening of the absorption lines or the veiling of
the spectrum of one of the components at the elongation where the
velocities of the star are of recession, are also indicative of the
existence of a gaseous stream in the system.

Table 5.1

PHASE-DEPENDENT INTENSITY VARIATIONS IN THE ABSORPTION

LINES OF SPECTROSCOPIC BINARIES

System	Observer and year	Negative results
μ^1 Scorpii	Bailey, 1896; Cannon, 1897; Maury, 1920	Struve and Elvey, 1942
V Puppis	Cannon, 1915; Maury, 1920; Frieboes, 1962	Popper, 1943
ζ Centauri	Maury, 1922	Popper, 1943
α Virginis	Struve, 1934	
σ Aquilae	Struve, 1937; Petrie	
β Scorpii	Struve, 1937	
2 Laceitae	Petrie	
29 UW Canis Majoris		
π Scorpii		
HD 47129	Struve, 1948	
AO Cassiopeiae	Struve and Horak; 1949	
W Ursae Majoris systems in general		

2.3,   In his review Chapter in Stars and Stellar Systems, Sahade (1960)
mentioned that another type of evidence for the existence of a flow of
gaseous matter is provided by emission lines that yield velocity curves
out of phase relative to either component of the binary. These velocity curves can be interpreted in terms of velocity (their K) and direction (relative to the radius vector joining the two stars, for instance) of a gaseous stream. Thus, from the relatively narrow H_α emission which, in the spectrum of HD 47129 (Struve, Sahade and Huang, 1958)
appears superimposed upon a broad emission feature, one derives a velocity distribution which suggests that the gaseous stream is strongly
deflected by radiation pressure from the O component of the system in
a direction which is nearly perpendicular to the radius vector joining

the two components of the system and points towards the observer at about phase 0.75 P; the velocity of such a stream would be some 350-400 km sec^{-1}. A similar relatively narrow emission again superimposed upon a broader emission is displayed by HeII λ 4686 in the Wolf-Rayet binary V444 Cygni (Sahade, 1958b). The behavior of such an emission velocity-wise appears to suggest a gaseous stream from the Wolf-Rayet component towards the O companion with a velocity of some 700 km sec^{-1}.

In the case of HDE 226868, the optical component of Cygnus X-1, the HeII emission appears to originate (Smith et al., 1973; Hutchings et al., 1973) in a gaseous stream that goes from the supergiant, "optical", component towards the X-ray companion. In fact, the radial velocities from the HeII emission suggest a velocity curve phase-shifted some 120° with respect to the velocity curve of the super-giant star and a stream velocity of the order of 150 (sin i)$^{-1}$ km sec^{-1} (cf. Sahade, 1976a).

3.   EVIDENCE FOR CIRCUMSTELLAR ENVELOPES

3.1.   The first observational evidence that led to the concept of the existence of circumstellar envelopes around components of close binary systems, came from the investigation of RW Tauri (Joy, 1942). Joy found that, as was mentioned in the introductory Chapter, the double H emissions that were present in the spectrum and flanked the corresponding stellar absorptions, underwent eclipses at the time when the large subgiant secondary passes in front of the primary component of the system. At second contact only the red emission was present, at third contact only the violet emission was seen, while at mid-eclipse the emissions were absent.

Joy interpreted these observations by postulating that the B9 primary was surrounded by a gaseous ring. Struve found similar behavior in SX Cassiopeiae (Struve, 1944b) and in RX Cassiopeiae (1944c) and later other eclipsing systems were found to display also the spectral features that are connected with a gaseous ring (cf. Sahade, 1960; Batten, 1973c).

The gaseous rings rotate with velocities of several hundred kilo-meters per second - around 350 km sec^{-1} in the case of RW Tauri - in the same direction of the orbital motion and the emissions that are associated with them are strong enough to be observed throughout the whole orbital cycle only in systems with relatively long periods. Otherwise, the emissions are detected only during eclipse when the exposure times become longer.

3.2.   The existence of circumstellar envelopes is not always indicated by the presence of double emissions that undergo eclipses. Sometimes the geometry of the system relative to the observer is such that the emissions are double but no eclipses occur, and many times the emissions are single; the radial velocities derived from them suggest that they reflect the orbital motion of the star which is surrounded by the envelope. Let us list the relevant cases, namely,

i)   the Wolf-Rayet binaries, where the Wolf-Rayet component displays Gaussian-like and flat-top emissions, the velocity curve of the object being derived from the measurements of the former type profiles;

ii) the eruptive variables, which display emissions associated
with the blue component;

iii) a system like V453 Scorpii, where the spectrum of the smaller,
more massive, component displays only emission lines;

iv) a system like β Lyrae where the emission is at least partly
associated with the secondary component. This statement was true at
the time of Belopolsky's (1893; 1897a; 1897b) and Curtiss'(1912)
radial velocity measurements of the emission at H$_\beta$, which seemed to
reflect the orbital motion of the companion to the B8 II star, and
is supported at present by a) the Balmer decrement and the trend of
the radial velocities that characterize the broad emission component
(Batten and Sahade, 1973) that partly describes the observed emission
profiles (Sahade, 1966; Batten and Sahade, 1973) that are present in
the spectrum of the star, and b) by the behavior of the CIV and perhaps
SiIV emissions in the ultraviolet (Hack et al., 1976a) which suggests
that they originate in matter around the secondary component of the
system.

In the case of KU Cygni and RZ Ophiuchi, the gaseous rings around
the F primaries produce an ultraviolet continuum which extends strongly
to longer wavelengths (cf. Popper, 1973).

3.2.1. In the introductory Chapter we referred to the puzzle posed
by β Lyrae and ε Aurigae because of the photographic depths of principal minima and the fact that even at mid-primary eclipse the lines
of the primary components are present in the spectra, while no trace
of the secondary stars is detected. The solution of the puzzle was
suggested by Huang (1963; 1965) who postulated that the circumstellar
envelopes around the secondary components have the characteristics of
a flat, opaque disk. At least in the case of β Lyrae, the density
of the disk must thin out toward the edges if the interpretation of
the "satellite lines" advanced by Batten and Sahade (1973), in the
sense that they arise from absorption of the radiation of the B8 II
component through the matter in the disk, is correct. It may even
be that the density thins out in all directions and the envelope has
an overall "spherical" distribution.

Disks surrounding one of the components in close binary systems
appear to be the rule whenever such components are compact objects
and there is a matter flow from the companion star. This is so in
eruptive variables and in X-ray binaries.

Indeed the blue component towards which the stream is directed in
eruptive variables, is a highly evolved object below the main sequence,
a hot subdwarf, definitely a white dwarf in the case of novae. The
emission lines that appear in the spectrum of the eruptive variables
and yield radial velocities that reflect the orbital motion of such
blue component, as indicated in 3.2, arise in the disk - the density
of which must be high - that surrounds the star.

3.2.2. The photometric observations of the eruptive variables produce
light curves that are quite unlike those that characterize eclipsing
systems, in the sense that they display a distinct maximum that lasts
for about half the orbital period and have led to the concept of a
"hot spot". These hot spots undergo eclipses of such a length that,
because of the relative dimensions of the stellar components as compared to the critical equipotential surfaces, must be located at the
outer edge of the disk that is around the blue component (Warner and
Nather, 1971; Smak, 1971) and, therefore, result from the heating of

the region of the disk where the collision with the stream coming from the larger companion takes place. We can consider a "hot spot" as an additional element in the gaseous structure in close binary systems whenever we deal with compact components and disks. According to Warner (1971) the bright spots will radiate maximum energy in the 0.1 keV region.

3.2.3. If we are concerned with even more compact components than those that are the case in eruptive variables, the matter flowing towards them will be highly accelerated and the kinetic energy that will be acquired will result in larger heating temperatures at impact than those that are relevant in the cases discussed under 3.2.2. The result will be a hot plasma at or near the stellar surface reaching temperatures of the order of 10^7 degrees or more and, therefore, emitting X-radiation at energies much higher than those that are characteristic of the "hot spots" in eruptive variables. We thus have an X-ray binary.

Therefore, the existence of X-ray binaries is also an additional evidence for the existence of gaseous streams and circumstellar envelopes in interacting binaries. We shall discuss them separately in Chapter 8.

3.3   Another evidence for the presence of circumstellar envelopes around components of close binaries is provided by polarization measurements. In the case of β Lyrae, the binary for which the variation of polarization with time has been most thoroughly investigated, these measurements are interpreted as suggesting electron scattering in the disk around the secondary component (Shakhovskoj, 1962, 1964; Shulov, 1967) and absorption of starlight by hydrogen before and after electron scattering (Appenzeller and Hiltner, 1967). In a review paper on polarization of variable stars, Serkowski (1971) mentions a few of the close binaries where changes in polarization during eclipse have been detected and states that he has concluded from his study of 14 bright southern eclipsing objects that "only the eclipsing binaries with emission lines in their spectra would show changes in polarization with time". From his observing list only TT Hydrae and V453 Scorpii did show variable polarization.

V444 Cygni, the Wolf-Rayet binary, is another example of variable polarization (Shulov, 1966; Hiltner and Mook, 1966). In fact, electron scattering is dominant in the envelope around the Wolf-Rayet component of the binary (cf. Kron and Gordon, 1950) as is shown by the fact that the photometric behavior of the primary eclipse - which is twice as wide as the secondary eclipse - is wavelength-independent.

Struve has interpreted as effects of electron scattering by the stream in β Lyrae the fact that the stellar absorption lines become broad and hazy during primary eclipse - an effect which is more pronounced after mid-eclipse - (Struve, 1957) and the changes in the radial velocity that he found immediately after mid-of primary eclipse in different cycles (Struve, 1958) of the order of some 10 km sec^{-1}.

4.   EVIDENCE FOR OUTER ENVELOPES

4.1.   The spectrum of a number of close binaries display absorption lines that suggest that they are formed in a medium where the radiation is diluted, that is, that they are formed at a distance of several stellar radii from the source of the radiation, in layers where electron densities are of the order of 10^{11} cm^{-3}.

In the case of β Lyrae, υ Sagittarii, HD 47129, the symbiotic star AG Pegasi and the Wolf-Rayet binaries, the effect is shown in HeI, the lines of the triplet series being the strongest, particularly those that arise from the metastable level 2^3S, namely λ 3888 and λ 10830. These lines suggest large velocities of approach and the values derived from them are of the order of -170 km sec^{-1} in β Lyrae (Sahade et al., 1959), -300 km sec^{-1} in υ Sagittarii (Sahade and Albano, 1970), -700 km sec^{-1} in HD 47129 (Struve, Sahade and Huang, 1958), -500 (perhaps even -1000) km sec^{-1} in AG Pegasi (López and Sahade, 1969), and over -1000 km sec^{-1} in the Wolf-Rayet binaries [-1300 km sec^{-1} in γ$_2$ Velorum (Sahade, 1955)]. In β Lyrae, γ$_2$ Velorum and AG Pegasi the lines display multiple components with intensities and profiles that vary with phase, indicative that we observe different layers of the "outer envelope" and that the densities are not uniformly distributed around the systems. An extreme case is provided by HD 47129 where the diluted lines are only present at the phases at which the velocities of the primary star are of recession, a fact which is consistent with the deflection of the gaseous stream from the secondary as discussed in 2.3.

In the case of W Serpentis, where the star involved is of a later spectral type than in the systems just mentioned, the presence of diluted radiation and, therefore, of an outer, tenuous envelope is suggested by the fact that the spectral features that dominate are those of a shell and the lines of FeI arising from non-metastable levels are either extremely weak or undetectable (Sahade and Struve, 1957).

4.2. In some systems the evidence for an outer envelope is provided by the presence in their spectra of relatively broad H emission with a steep Balmer decrement that is strong in H$_\alpha$ and already very weak or absent at H$_\beta$. Such an emission must arise in a tenuous envelope around the system. This is so in HD 698 (Sahade, 1966), U Coronae Borealis (Struve, Sahade and Huang, 1957), AO Cassiopeiae (Struve and Sahade, 1958), and it is also true in β Lyrae as far as the relatively narrow emission component that partly describes the observed profiles (Sahade, 1966; Batten and Sahade, 1973). The combined analysis of the ultraviolet spectral scans and photometry also suggests the presence of an outer envelope where the FeIII in emission should originate (Hack et al., 1976a).

In the Wolf-Rayet binaries the flat-top emissions - typical of uniformly expanding spherical atmospheres (cf. Underhill, 1969) - arise from the layers of the outer expanding envelope where the absorption lines suggesting the presence of diluted radiation are formed. Actually the flat-top profiles have usually a P Cygni absorption that yields the large velocity of approach that we have mentioned in the case of HeI λ 3888.

4.3. A number of systems display forbidden emissions in their spectrum which suggests that we are getting information from very tenuous regions of the outer envelopes. Table 5.2 lists five systems where we observe forbidden transitions of the elements that are indicated.

Table 5.2.

SYSTEMS WITH FORBIDDEN EMISSIONS

Object	Elements present
RY Scuti	[FeIII]
W Serpentis	[FeII]
υ Sagittarii	[CaII] , [FeII]
T Coronae Borealis	[NeIII] , [OIII]
VV Cephei	[FeII], [SII], [NiIII], [CrII]?

We could add the symbiotic stars where the forbidden emission suggest that we observe layers with electron densities of the order of 10^6 cm^{-3}.

Rocket observations of γ_2 Velorum have permitted the detection of the intercombination line CIII] λ 1909 (Stecher, 1968), which should a-rise from layers close to those where the lines affected by dilution are formed. This line suggests a velocity similar to the one which is derived from HeI λ 3888 and HeI λ 10830. In this case we are dealing with layers with electron densities of the order of 10^{10} cm^{-3}.

5.   PHOTOMETIC EVIDENCES FOR THE GASEOUS STRUCTURE

When discussing the evidence for the three elements of the gaseous structure in close binaries purposedly we have made no reference to photometric evidences. Actually, we should refer the reader to Wood's (1957) paper already mentioned at the beginning of the present Chapter and to Batten's (1973) book. There are effects that appear in the light curves and seem to be connected with the fact that mass is being shed by one of the components of the system. Period changes, that is, decreasing or increasing periods, and erratic fluctuations and abrupt variations of the period, are indicative of mass loss as first suggested by Wood (1950) and lately explained by mass exchange through a proposal by Biermann and Hall (1973). There are other causes that result in changes of period and the whole problem of period changes will be dealt with in Chapter 6.

The other effects like "humps", asymmetries in the two branches of minimum, variations in light depth, changes in shape of the light curve around mid-eclipse and at other phase intervals, are most prob-ably connected with the phenomena that occur in interacting binaries Unfortunately there are few spectrographic and photometric observations carried out simultaneously to lead to more definite conclusions. In the case of β Lyrae, the star that has been the subject of a larger number of investigations, Struve and Wade (1960) established that the

significant changes in radial velocity that are found in the partial
phases of the eclipse, as discussed in 2.3., correlated with systematic differences in the light curve, in the sense that the integrated
brightness of β Lyrae was lower when the velocities were more positive
i.e., less distorted. This, on the otherhand, correlated with
the strengthening of the lines from the outer envelope. More information of this kind is certainly needed over extended time intervals.
An important present example is U Cephei, which will be discussed in
some detail in Chapter 10.

6.   RADIO EMISSION IN CLOSE BINARIES

One of the most important developments in the last years in the
field we are dealing with in the book is the detection of radio emission in a number of close binaries. The list includes:

some X-ray doubles like Scorpius X-1 (Andrew and Purton, 1968;
Ables, 1969), HDE 226868 - Cygnus X-1 - (Hjellming and Wade, 1971;
Braes and Miley, 1971; Hjellming et al., 1971), Cygnus X-3 (Braes
and Miley, 1972; Gregory et al., 1972a) and Cygnus X-2 (cf. Hjellming,
1976c),

"classical" close pairs like β Persei and β Lyrae (Wade and
Hjellming, 1972), CC Cassiopeiae (Gibson and Hjellming, 1974),

the RS Canum Venaticorum binaries AR Lacertae (Hjellming and
Blankenship, 1973), UX Arietis (Gibson et al., 1975a), RT Lacertae
Gibson et al., 1975b), and HR 1099 (Owen et al., 1976),

the Wolf-Rayet binaries γ$_2$ Velorum (Seaquist, 1976a), and HD
193793 (Florkowski and Gottesmán, 1976, 1977).

In addition, the symbiotic objects AG Pegasi (Gregory et al.,
1977), V1016 Cygni (Purton et al., 1973; Altenhoff et al., 1973;
Seaquist and Gregory, 1973), RY Scuti (Hughes and Woodsworth, 1973;
Hjellming et al., 1973); RX Puppis (dos Reis, 1975; Seaquist, 1976b),
perhaps SY Muscae (dos Reis, 1975, 1976),

the BQ[ ]   star MWC 349 (Braes et al., 1972; Altenhoff and
Wendker, 1973; Baldwin et al., 1973), and Nova Delphini 1967
and Nova Serpentis 1970 (Hjellming and Wade, 1970) and Nova Scuti 1970
(Herrero et al,,1971), are also radio sources.

The radio emission appears to be variable, for instance, in β
Lyrae and in HD 193793 (Florkowski, 1977) and strongly variable in
systems like β Persei (Algol), AR Lacertae and Cygnus X-3. Actually,
"variable radio emission, mainly at the higher GHz-range frequencies"
appears to be one of the characteristics of the "classical" binaries
that we have listed (Hjellming, 1976c), the time scale of variability
being from minutes to hours. In the case of Algol, the object which
has been followed more extensively in its radio behavior, such a behavior was found by Hjellming et al. (1972) to be predominantly thermal
and to be characterized by being "erratic, with ocassional periods
of strong flaring and long quiescent" intervals in between. They
concluded that the radio emission source was within the volume of the
equipotential lobes and proposed that the component that ejected
mass undergo "starquakes", once every 25 years or so, as a mechanism
to produce discontinuous ejections and to energize "a large and very
hot thermal plasma" which "would be the basic source of the 'typical'

radio flare". However, a strong radio outburst observed in 1975
through very long base interferometry at 7859 MHz (Clark et al., 1976)
suggests that its origin may be non-thermal as it seemed to be also
true of the event on July 11, 1972, observed by Hjellming et al. (1972)
at 2695 and 8085 MHz. There seems to exist a correlation between the
radio flares and the optical spectrum of Algol, in the sense that
when the radio spectral index, which decreases during flares, becomes
smaller than about -0.3, weak emission does appear on the red wing
of the CaII-K line (Bolton, 1972, 1973). According to Hjellming (1976c)
"the most fruitful working hypothesis for explaining the Algol radio
emission, and presumably that of similar radio binaries is that of
incoherent synchrotron radiation from relativistic electrons in a
relatively dense plasma environment".

Gibson and Hjellming (1974) after analyzing the information on six
variable radio binaries, believed that it is now "nearly ruled out
the possibility for a thermal interpretation of the Algol-type radio
events" and suggest that "the source of energy supply for the processes that eventually result in variable radio emission, is the gravitational energization of matter falling in or near the surface of
one of the binary components".

Woodsworth and Hughes (1976) have proposed a model for Algol where
two sources for the radio emission are recognized. One of these is
thermal and arises from an optically thin HII region which surrounds
the optical object,"while the other is a non-thermal flare component
probably associated with the region of mass exchange".

The RS Canum Venaticorum stars to which we specially refer to in
Chapter 9, appear to have the general property of being radio emitters.
In the particular case of HR 1099, the radio emission may be circularly
polarized which suggests a synchrotron-type process at work and, therefore, the presence of a magnetic field.

In the case of Cygnus X-3, an outburst observed early in September,
1972, at 10522, 8085, 6630 and 2695 MHz suggested, on account of its
spectrum, the nature of the decay and the measured linear polarization,
that it was of non-thermal origin. It actually displayed the characteristics that are "expected for synchrotron radiation from an expanding cloud of relativistic electrons" (Gregory et al., 1972b). This
conclusion was again true for the "highly-polarized event of May, 1974"
(Seaquist et al., 1974) and may possibly hold for every flare activity
of Cygnus X-3. The interpretation of the radio behavior in the quiet
phase is less understood as there are "slow variations in the mean
level for each day and modulations on each and every day" (cf. Hjellming,
1976a), which may be related to phenomena connected to the "ejection
and acceleration of the relativistic particles" according to Hjellming
et al. (1974). These authors believe that "other aspects of the pro-
longed injection of particles into the radio sources" may be responsible
for "the apparent exponential decay which, in the major events, precedes the power-law decay of the flux densities expected on the basis
of the simplest models".

As far as Scorpius X-1 is concerned, there are three radio sources
associated with the system (Hjellming and Wade, 1971b), the one in
the middle coinciding within a few seconds of arc with the X-ray source.
The three radio sources suggest that the radiation is non-thermal. Coordinated campaigns for simultaneous X-ray, optical and radio observations of Scorpius X-1 carried out in 1971 and 1972 have failed to
detect any correlation between the X-ray/optical activity and the radio

flaring of the object (Bradt et al, 1975; Canizares et al., 1975).
However, in May-June 1975 (Hjellming, 1976b) the radio decay "was
qualitatively similar to the decay seen at X-ray wavelengths during
this period".

With only a small number of cases at hand and a very small amount
of information about the behavior of the radio emission it is still
too early to draw conclusions. However, it seems that such a radiation is related to the gaseous structure of the systems and carries
a message that could tell something about the physics of the mass loss
in components of close binaries - at least at certain stages of their
evolution. It seems probable that all X-ray binaries should be radio
emitters, but is is not yet clear in which other systems we should
expect to detect radio emission. However, it may prove to be true
that all systems with gaseous structure and mass flow from one of
the components are in greater or smaller degree radio sources.
Florkowski (1977) believes that, for radio observations, interacting
binaries with gaseous streams and disks or rings, phenomena which,
as we mentioned earlier, appear to be confined to the orbital plane,
are indeed affected by geometrical selection effects, as well as by
optical depth or frequency selection effects.

7.   GENERAL COMMENTS

7.1. We have described the gaseous structure that is recognized in
interacting binaries and given the types of observational evidence
that tell us of its existence.

The behavior of the emission lines that suggest the existence of
gaseous rings around components of close binaries also suggest that
the motions in the rings are complicated. This not only results
from the variations in character and intensity of those lines but
also from the distribution of the velocities derived from the absorption features. In the case of RZ Scuti (Hansen and McNamara, 1959)
a step-like velocity curve was derived and an interpretation in terms
of the disappearance and reappearance of parts of the gaseous formation from behind the primary star was advanced. In the case of
massive systems like HD 47129, AO Cassiopeiae, V448 Cygni and β Scorpii,
the velocities from the lines of the "secondary" component have a distribution which may be different in different cycles and suggest that
they do not reflect purely orbital motion (Sahade, 1962).

Therefore what we have called circumstellar envelopes may have very
different structure and density in different systems. The envelopes
in β Lyrae, in ε Aurigae and in eruptive variables are good cases to
bring up again as quite different in density as the gaseous rings in
systems like RW Tauri and S Velorum. As for the outer envelopes we
had concluded that the matter in them was not homogeneously distributed and their density varied with phase and at different times.

Since the material in the two envelopes is supplied by the flow of
the matter which is being shed by one of the components of the system
through what we have called the "gaseous stream", variations within
the envelopes and variations of the envelopes with time should be
ascribed to the anisotropic character of the stream and to variations
of the stream with time.

The "flickering" which characterizes the radiation from the "hot

spot" in eruptive variables (cf. Warner, 1971a) may be giving us information on the fluctuations in the colliding stream in those objects.

7.2.   In certain stages of evolution of a close pair the physical
characteristics and the behavior of the envelopes may be such that the
stellar features could be strongly masked and we could find ourselves
with a very peculiar object, the binary nature of which may be quite
difficult to disentangle.

In some cases this could be done after collecting enough material
and analyzing it in the light of the possibilities as far as the
gaseous structure of the system goes. During the past years several
shell stars have been found to be binaries (among others by Harmenec
and coworkers) and an object like HR 2142 (cf. Peters, 1973), a very
broad-line B1Vnne star with periodic shell formation, has been understood in terms of an interacting close double. Actually it is now
believed that all shell stars may be binaries and, furthermore, the
question has been raised as to whether some Be stars could also be
explained in similar terms.

An I.A.U. Symposium that was held late in 1975 (Slettebak, 1976)
was devoted to this problem and the feeling was that it was too early
to accept a general explanation in terms of binary nature of the objects. More Be and shell stars have to be analyzed and proved to be
binaries before this could happen.

In this context, perhaps we should mention again W Serpentis. In
3.1 we have stated that its spectrum is dominated by shell features
and even a line like Mg II λ 4481 which looks stellar and at one time
it was thought that it could provide the star's velocity curve (Sahade
and Struve, 1957), is blended with emission features (Greenstein et
al., 1970). The velocity curves that are derived suggest some kind
of orbital motion but the scatter is very large; distortions must
affect them and they could not provide information on the orbital
elements of the system. The binary nature of the object is largely
suggested by the light curves which is of the eclipsing type; the
brightness of W Serpentis between principal eclipses is, however,
quite atypical, in the sense that the variation is continuous and
there is no distinct secondary minimum but what we can describe as
three maxima (cf. Walraven, 1969). The gaseous structure in the system must be responsible for such a behavior, and the same should be
true with HD 187399 (Hutchings, 1973). These cases are, therefore,
relatively simple to be interpreted in terms of binary systems; only
now more complicated cases are beginning to be understood in similar
terms.

We should also mention in the present context, that in the eruptive variables, the relative light contributions of the stars, the
disk and the hot spot define the shape of the light curve that a
particular object will have. In U Geminorum the eclipse is due to a
total eclipse of the bright spot and a partial eclipse of the disk,
and in UX Ursae Majoris the bright spot is slightly fainter than
the blue star and both are eclipsed (cf. Warner, 1971a).

7.3 For the sake of completeness, perhaps we should make brief mention of the question of the differences of the values of the systemic
velocity, γ, as derived from the absorption lines of each component,
the question of the erratic variations in radial velocity displayed
by one of the components, and the question of mass outflow through
external Lagrangian points.

7.3.1.   The difference in the γ's for each component has been interpreted by Sahade (1959, 1962) as resulting from the fact that the
lines of one of the components originate in an expanding atmosphere
that arises from shell-like ejection of matter from the star.   Thus,
the star that would be ejecting matter will give the more negative
of the γ's.   This interpretation has been applied to 29 UW Canis
Majoris, HD 47129, AO Cassiopeiae, V448 Cygni, β Scorpii and α Virginis (Sahade, 1962).

   The case of the Wolf-Rayet binaries where the velocity curves from
different emission lines suggest a different systemic velocity is
altogether different because these differences probably arise from
problems related with the wavelengths and/or the line profiles, or
to perhaps other causes.

7.3.2.   Erratic radial velocities were found in systems like UX Monocerotis (Struve, 1947) and S Velorum (Sahade, 1952), with changes, in
the former case, of some 100 km sec^{-1} in a five-hour interval.   Actually nearly all systems with gaseous streams show a large scatter in
the radial velocities but such changes are perhaps not as drastic.
This behavior may be linked with instabilities connected with the
process of mass loss.

7.3.3.   In systems like β Lyrae and SX Cassiopeiae there is evidence
for streaming of matter through the outer Lagrangian points.   In the
case of β Lyrae the velocities from H and HeI (Sahade et al., 1959),
at about secondary minimum, are more negative than those from other
elements.   Likewise, in SX Cassiopeiae (Struve, 1944b) the conclusion
aimed at accounting for the velocity distribution in the phase intervals that correspond to about secondary minimum.

CHAPTER 6

CHANGES IN PERIOD

1. GENERAL REMARKS

The period of orbital revolution of the two stars in a binary system is one of the fundamental parameters of such a system and is normally the one which can be determined with the highest precision - usually many orders of magnitude better than the other elements. In eclipsing binaries, this is almost always determined from observations of times of minimum light, which are assumed to correspond to the times of conjunction. The "light elements" usually consist of one well determined time of minimum, the primary epoch, E_0, given in Julian days and the period, P, expressed in days and decimals of a day. Thus, a predicted minimum can be computed for any desired epoch E and compared with the time actually observed by means of the expression:

$$\text{Pr. Min.} = \text{\Moon}_{\odot} + P \cdot E \qquad (1)$$

Period studies are generally conducted by plotting the (O-C)'s - the differences in time between the observed minima and those computed from the light elements - as a function of the epoch itself.

The determination of the light elements - in particular the period - from a series of observed minima recorded, by necessity, at irregular intervals, is not always an easy matter. Occasionally multiples or submultiples of the period will fit the observations equally well. Procedures for deriving the light elements have been adequately described elsewhere (e.g., Batten, 1973a). Occasionally, radial velocity observations show that the original photometrically derived period is either twice or one-half the true value.

For two point masses, revolving in orbits determined by their mutal gravitational attraction, the period should remain constant. In the case of actual stars, of course, there should be exceedingly small changes as mass is converted into energy and radiated into space, but these should be so minute as to be nearly undetectable observationally in the lifetime of the star. In many cases, however, we observe changes in the period which can be either increases or decreases. When secondary minima are deep enough to permit accurate determinations

55

of their times of occurrence, they sometimes show changes corresponding
to those of the primary and sometimes show changes in the opposite
direction. In general, the period changes may be divided into systems
in which the variations themselves are periodic and those in which
they are not. The periodic group many be further subdivided into
those in which changes in primary and secondary agree in size and direction (lengthening or shortening) and those in which the change in
direction is in the opposite sense.

   In the first of these subclasses, the changes are believed to be
caused by variations in the time of travel of the light from the system to us as the pair moves around its common center of mass with a
third body in the system. A well known case is that of Algol. In
some cases, the observations would require a highly improbable value
of the mass function of the three body system; this at least suggests
that further observations will show that the supposed periodicity is
not real. For completeness, we should mention that a continuous decrease of period could be caused by a system passing through a dense
cloud or even by currents in circumstellar material surrounding the
system, but no such case is definitely known.

   When the plot of the secondary minima shows instead a sinusoidal
variation 180° out of phase with those of the primary, the conventional explanation is a rotation of the line of apsides in an eccentric orbit. Since the stars move in accord with Kepler's second law
(the law of areas), and since in a elliptical orbit the distance actually travelled in the relative orbit between primary and secondary
eclipses generally will not be the same as that between secondary
and primary, the observed position of secondary relative to adjacent
primary minima will vary with the changing orientation of the line
of apsides relative to the line of sight to the earth. The result
will be variation in observed times of primary and secondary as just
described. Well documented cases of this type are relatively rare.
The orbits must be appreciably eccentric - something of a rarity among
close systems -, the secondary should be reasonably deep, and the observations must cover a considerable fraction of the period of apsidal
rotation; this is normally a span of many years. In a sample of 152
close binaries, selected because of large observed period changes,
Rafert (1977) found 4 which showed pronounced apsidal motion.

   Finally, there are many cases in which the period varies in either
direction and apparently at irregular - certainly at unpredictable -
intervals. It is now fairly well accepted that these are caused by
mass ejection from one of the components.

   For completeness we should mention that theoretical calculations
of evolution of close binaries assuming no loss to the entire system
of either mass or angular momentum, predict a gradual shortening and
then subsequent lengthening of the period as mass is transferred from
one component to the other.

2. LIGHT TIME CHANGES

   If the center of mass of the binary system is moving in an orbit
around the center of gravity lying between the system itself and a
relatively distant third component, the observed period will be longer
than the true one when the system is receding from the earth and

shorter when it is approaching simply because of light time effects. That is, when the system is receding the observed interval between successive eclipses will be longer than the true period because of the greater distance between the system and the earth at the time of later eclipse; the reverse will be true when the system is approaching. If the orbit of the eclipsing pair around the center of mass is nearly circular, the plot of (0-C) versus epoch will approximate a sine curve. For an elliptical orbit, the curve will be distorted and in principle the eccentricity of the orbit and its orientation relative to the earth can be determined from the curve. If secondary minima are deep enough for a similar plot to be made, it should of course be identical to that of primary to within observational scatter.

Known cases of this sort are rare. Observations of times of minima over many years are required and if there are other relatively sudden causes of period change, such as those to be discussed later, they may make difficult an evaluation of the periodic variations.

As a well known example, we cite the treatment by Dugan (1937) of the system RT Persei, P = 0.85 days. Using all available minima, he found the residuals were beautifully represented by a light-time orbit whose period, P', was 37.2 years. This study was unique in that it included observations spanning more than thirty years made by the same observer (Dugan) using, except for one brief interlude, the same telescope and photometer, and the same methodology for determining times of minima. The importance of this will be appreciated particularly by those who have attempted period studies using observations from various observers, especially those made before the days of photoelectric photometry. The importance of observing times of secondary minimum was also shown by this study. Although in this system they were determined with far less precision than those of primary, they were sufficiently accurate to show that their (0-C) residuals followed very closely those of primary and that therefore a revolution of the line of apsides was not involved in any significant fashion.

The observed minima up to 1937 thus showed clearly a periodic variation which was well explained by the hypothesis of motion in a three body system. Yet beginning about 1940, the observed minima showed systematic deviations from the predictions. Dugan himself had remarked that period changes frequently occur immediately after publication of a careful study.

Another effort was made by Vasileva (1952) to explain the variation of RT Persei by a combination of periodic terms, to represent the later observations. Frieboes-Conde and Herczeg (1973) have rediscussed the available material. They find a thoroughly satisfactory representation of all available minima if, in addition to the light time effects, we postulate a relatively sudden decrease of 0.4 sec occurring somewhere around JD 241 9550. This leads them to a light-time orbit with a period P', of 38.6 years, a_{12} sin i of 2.95 A.U. (a_{12} is the semi-major axis of the light time orbit) and an eccentricity of 0.1. The mass of the hypothetical third body, computed for widely different assumed values of the inclination and realistic limiting values for the mass of the eclipsing pair, ranged from $0.31\mathfrak{M}_\odot$ to $1.10\mathfrak{M}_\odot$. In another discussion using observed manima from 1905 to 1973, Ahnert (1974) found a regular variation with a period of 37.2 years upon which irregular variations were superimposed. He computed a value of $0.53\mathfrak{M}_\odot$ for the third component. These are all reasonable values; moreover they suggest that this is one system in which it might be worthwhile to look for "third

light" in interpretation of the light curve. Dugan had indeed discussed his visual observations with this possibility in mind, but
found no significant changes in the computed elements derived from
them; whether the same conclusion would apply to multicolor photoelectric work is another question.

Figure 6.1 shows an (O-C) versus epoch plot for RT Persei through
1976, as collected by Rafert (1977).

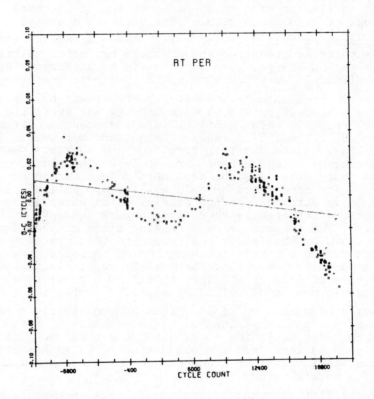

Fig. 6.1. Results of period variations in the system,
RT Persei.

Frieboes-Conde and Herczeg discussed fourteen eclipsing systems
(including RT Persei, but not Algol) for possible light time effects.
They feel that in none of them do the observations "prove definitely"
the existence of a third body, but that seven should be kept on the
list of possible candidates. Their discussion omits (as they point
out) a number of interesting systems either because separate studies
are in progress or because nothing could be added to earlier published
discussions.

Sometimes the attempt to explain observed changes in period has
gone to the length of invoking a fourth and even a fifth body in the
system. However, these have never stood the test of continued

observation or of correlation with other observed properties. A well
known case of this sort of treatment is found in the history of the
study of Algol; it is described in some detail in the discussion of
this system in Chapter 10.

3.   ROTATION OF THE APSE

It is well known that any of three effects can cause a rotation
of the line of apsides of a binary system. These are: (1) the gravitational influence of a third body; (2) rotation of the apse required by the general theory of relativity, and (3) perturbations
caused by the distorted shapes of the components. In this section
we shall be concerned almost exclusively with the third effect, which
in almost all cases is by far the largest. In his book Batten has
given a thorough discussion of the physical cause, which will not be
duplicated here. Semeniuk (1968) has also given a comprehensive
study and Herczeg (1975) has summarized recent work.

There is a direct relation between the gravitational field of a
non-spherical body and the internal density concentration in that
body. Russell (1928) apparently was the first to realize that knowledge of the density distribution in stellar interiors could be ach-
ieved by measuring the rate of advance of the line of apsides. He
assumed that the stars are ellipsoids with their major axes along
the line joining the stars. Walter (1933) assumed the stars were
rigid bodies; then Russell's assumption would not be true because
of varying velocities in elliptical orbits. Walter predicted a recession of the apsidal line. Kopal (1938) criticized Walter's work
and claimed to substantiate Russell's conclusion. However, Cowling
(1938) pointed out the Kopal's discussion was based on lack of understanding of Walter's work and contained at least one serious basic
error. The weakness in Walter's conclusion is his assumption of
complete rigidity of the stars. Of various treatments of the problem
we select Cowling's as most suitable for a brief summary.

Cowling pointed out that the time needed for a star to adjust to
changes in an external gravitational field is essentially the period
of its free adiabatic oscillations and this is much smaller than the
orbital period of the system. Hence, the shape would be essentially
the equilibrium shape and thus Russell's basic assumption is justified. However, Russell failed to allow for the changing shapes of
the stars with varying separation in their elliptical orbits. The
more rigorous treatment by Cowling produced identical results except
for a change in one numerical constant.

Cowling confirmed by detailed calculation for the case of uniform
stars (incompressible fluids) the assumption that the shapes at any
instant closely approximate the equilibrium form and thus that the
stars could be represented by aligned ellipsoids. He found that ε
(the ratio of the period of orbital revolution to the period of rotation of the apse) could be expressed by:

$$\varepsilon = 0.75 \left\{ \left(1 + 16 \; \frac{\mathfrak{M}_2}{\mathfrak{M}_1} \right) \left(\frac{R_1}{R} \right)^5 + \left(1 + 16 \; \frac{\mathfrak{M}_1}{\mathfrak{M}_2} \right) \left(\frac{R_2}{R} \right)^5 \right\} \qquad (2)$$

where as usual $\mathfrak{M}_1$ and $\mathfrak{M}_2$ are the stellar masses, R_1 and R_2 the radii, and R the separation of their centers. The separation was assumed to be large compared to the stellar radii; fortunately in most sys-tems showing elliptical orbits the components are relatively well separated.   Equation (2) is identical with that derived by Russell except that Russell's coefficient preceding the mass ratios was less than half the above value.   The proof depended only on orders of magnitude and hence the conclusion should hold for any density dis-tribution.   Therefore while the factor 0.75 applied only to stars of uniform density, the expression could be generalized to allow for other cases.   On the assumption that the stars' angular velocities of rotation are equal to the mean orbital velocity, Cowling devised a formula containing terms in e^2 and e^4, where e as usual was the orbi-tal eccentricity.   Neglecting these, the expression reduced to:

$$\varepsilon = k_1 \left(\frac{R_1}{R} \right)^5 \left(1 + 16 \; \frac{\mathfrak{M}_2}{\mathfrak{M}_1} \right) + k_2 \left(\frac{R_2}{R} \right)^5 \left(1 + 16 \; \frac{\mathfrak{M}_1}{\mathfrak{M}_2} \right) \qquad (3)$$

Again, this agreed with Russell's formula but with factors of 16 in-stead of 7.   The k's are numerical constants which depend on the density distribution of the components.   The value ranges from 0.75 for a star of uniform density throughout - as in equation (2) - to zero for a point source with all the mass concentrated at the center. For polytropic models of index n,

$$k = \tfrac{1}{2} \left(\Delta_2 (n) - 1 \right). \qquad (4)$$

$\Delta_2 (n)$ is a function which has been tabulated by Chandrasekhar (1933).

Equation (3) gives in any case significantly higher ratios of central to mean density than does Russell's equation and values that are more in accord with theories of stellar interiors.

Certain comments on the use of equation (3) are in order.

1.   Note that Newtonian gravitational effects only are assumed.   A relativistic effect is also present.   In most systems, errors introduced

by neglect of this effect will be an order of magnitude or more below
those from other uncertainties, but if the other quantities are exceptionally well determined, they should be investigated. Neglect
will give too large a value for k and thus too low a degree of central
condensation. In some systems, it may actually prove the major factor -
e.g., DI Hercules (Rudkjobing, 1959).

2.   The effect of a third body can not only alter the motion of
the apse, but also because of the light time effect complicate the
problem of determining the true period of rotation from the observed
minimum.

3.   The system must have been observed long enough so that a reliable determination of the period of apsidal rotation can be made.
A relatively sudden change in period is sometimes impossible to distinguish from a portion of a sinusoidal variation. If secondary minimum has been well observed, this will be of considerable value. The
(0-C)'s should of course show the same amplitude but 180° phase difference from primary. Even if the secondary has been observed only
at one epoch, its displacement from mid-point on the light curve (and
in a few extremely favorable cases its width relative to primary) can
substantiate whether or not the observed changes in times of primary
are really due to apsidal motion. Fortunately, most systems having
eccentric orbits are sufficiently well separated so that sudden changes
in period are rare.

4.   The relative radii enter to the fifth power. For meaningful
results, the light curve must be of high quality and the system of a
nature to permit a determinate solution. Normally this means a total,
annular, or deep partial primary eclipse and freedom from large distortion or re-radiation effects.

5.   The mass ratio should be well determined. Normally, a two
spectrum radial velocity curve is required.

6.   If the eccentricity is large, the more precise formula given
by Cowling is needed. This contains terms in

$$(1-e^2)^3 \quad \text{and} \quad (1 + \tfrac{3}{2} e^2 + \tfrac{1}{8} e^4).$$

If neglect of these introduces errors comparable with those of the
other certainies involved, then the more complex treatment is required.

7.   Equation (3) contains two "unknowns", k_1 and k_2. If the stars
are similar, the assumption $k_1 = k_2$ is perhaps not unreasonable; otherwise, a weighted mean, usually designated $\bar{k}_2$ with weights determined
by the respective coefficients is perhaps the best that can be done.
In some cases (as for example when the radii are greatly different),
the quantities $(R_n/R)^5$ for one star may be small enough to be insigni-
ficant, and we can essentially determine the degree of condensation of
one component alone.

Figure 6.2 shows a plot by Rafert (1977) of the (0-C) diagram for
V526 Sagitarii; this is a beautiful example of the diametrically opposite behavior of primary and secondary minima.

Let us now turn to a few cases of practical application. One factor
which is now beginning to be exploited is the use of observations of
apsidal motion and the ratios of central to mean densities deduced there-
from, to study the evolutionary changes that occur because of the thermonuclear processes in the stars' interiors. The changes for an individual star would of course be many orders of magnitude smaller than

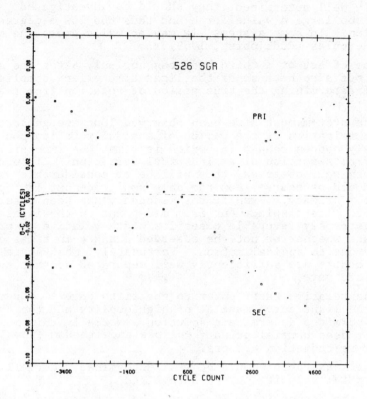

Fig. 6.2. Evidence of apsidal rotation in
V526 Sagitarii.

could be observed in any reasonable time span, but comparison of different systems in different evolutionary stages could yield informative
results. In particular, it might be possible to distinguish whether
a given system is in pre- or post-main sequence evolution.

A certain amount of work to this end has been done, and has been
summarized by Petty (1973) in his study of the problem. Kushawa (1957)
applied this test to his theoretical models as did Schwarzchild (1958)
to his. Wood (1965) computed the constant for the subgiant component
of TX Ursae Majoris. Kopal (1965) computed it for several observed
systems. Both Kopal's results and those of Schwarzschild showed systematic differences between the theoretical values and those computed
from the observations in the sense that the "observed" values of k
systematically fell below either the initial or the evolved time line.
Petty (1973) has summarized possible causes of the discrepancy.

1.   The theoretical models used were among the earliest computed
and modern models might fit better the observational data.

2.   The treatment of the observational data was not sufficiently
complete.

3.   The basic observational data were not sufficiently precise.

4.   A treatment emphasizing the evolutionary character of $\overline{k}_2$ (the
apsidal constant) is needed.

5.   A systematic difference between "observational" and "theoretical" models actually exists; this will require re-examination of the
basic assumptions.

Mathis (1967) tried to solve the problem by using later values of
opacities; he found better agreement but still a systematic difference.
He considered the observational determination of the relative radius
to be the major problem, but it is difficult to see how this could
cause the <u>systematic</u> differences.

Semeniuk and Paczynski (1968) made a detailed comparison using new
models running from $4\,\mathfrak{M}_\odot$ to $16\mathfrak{M}_\odot$. They found that the relation between
$\overline{k}_2$ and effective temperature was not very sensitive to chemical composition. They also found better agreement between theory and "observation" than did Schwarzschild or Kopal. However the computed coefficients, $\overline{k}_2$, were still systematically larger than the "observed"
ones. Other comparisons of models with "observed" values have been
made by Mathis and Odell (1973) and by Odell (1974). Batten (1973a)
has considered critically the observational determinations of $\overline{k}_2$ and
finds 11 worth tabulating; two years later Herczeg (1975) listed 13.
Of these, listed by Batten only two (Y Cygni and CO Lacertae) are
considered to be first class and two more (GL Carinae and AG Persei)
second class; the other 7 are rated third class. Herczeg listed four
as "definitive", 8 as "positive" and one as "?"; the definitive ones
were those in the top two classifications of Batten.

Petty's treatment of the subject used a series of stellar models
of mass $2\mathfrak{M}_\odot$, $5\mathfrak{M}_\odot$, $10\mathfrak{M}_\odot$, and $20\,\mathfrak{M}_\odot$ with the same initial compositional
mixture. His models begin where the star has just completed the Hay-
ashi phase of contraction and continue until the hydrogen in the core
is completely exhausted. He finds that comparison between theory and
"observation" shows that secular variation of $\overline{k}_2$ plays an important
role. For most of the cases studied, Petty believes the differences
between the two values of $\overline{k}_2$ can be reconciled in terms of the evolutionary history of the system.

Recently, Stothers (1974) has applied the apsidal motion test to
models of main sequence stars. He pointed out the importance of the
opacities used in determining the results of the computations. He
used the new "Thomas-Fermi" opacities to compute values of the apsidal
constant for both unevolved and evolved main sequence stars. The
agreement was slightly better but the values were still too large for
any reasonable assumed composition when compared with empirical constants for well observed systems. Computations assuming uniform rotation lowered the theoretical constants only slightly. However, consideration of normal evolution during core hydrogen burning produced
"good detailed agreement" with the observations. The comparisons were
made with Y Cygni, α Virginis, AG Persei, CO Lacertae, and CW Cephei.
Three of these were among Batten's "first class" and "second class"

systems and those which Herczeg considered "definitive"; for CW Cephei
Stothers had available recent unpublished observations of I. S. Nha.

Nha (1975) later discussed his observations. The importance of CW
Cephei is increased by the fact that it is a member of an association.
Nha finds an extremely short apsidal period of 39 years. Using the
diagram published by Semeniuk and Paczynski (1968), he was able to
infer the central hydrogen content and hence the age for three different assumed compositions. The values ranged from 23 to 42 x 10^6
years.

Nha's period study of the system emphasizes the importance of observations of times of minimum at various epochs. Even though his
own observations stretch over 8 years and are of high precision, it
would not be possible to put a great deal of trust in the derived
apsidal period were it not for three minima observed by Abrami and
Cester some 14 years earlier.

The problem is, of course, far from settled. Discussion of individual systems has been given both by Petty and by Batten, as well as
by Stothers. As in so much of astronomy, new and reliable observations
are needed. We feel that it is too early to hope that the age of a
system of given mass may be uniquely fixed by determination of its
apsidal constant, especially in view of the uncertainties in its determination as emphasized earlier in this chapter and also by Batten in
his discussion. Nevertheless, there is at least some hope for help
in discrimination when very young and much more envolved systems are
found in the same position on the H-R diagram.

4.   IRREGULAR PERIOD CHANGES

Changes in the periods of a few individual systems puzzled astronomers for many years and suggestions of various sorts - some of them
rather improbable, such as encounters with clouds of interstellar
material - were made. Complex light elements were frequently derived
having terms in E^2, E^3, $\sin(\omega + aE)$, or similar ones. The first
publication approaching in any way a general study seems to have been
made by Dugan and Wright in 1939 (Dugan and Wright, 1939). Working
from minima determined from patrol plates they studied 28 systems.
This paper appeared at a time when, judging from the number of publications on the topic, there was a good deal of interest in apsidal
rotation. In almost every individual system they discussed the possibility of the observed changes being due to apsidal motion was mentioned.
In no case, however, was this clearly established as the cause, and in
most instances the observed variation showed no periodicities or other
regularities. Dugan and Wright made no attempt to give a physical explanation for the changes.

The first effort to make a general study and also to suggest a possible physical explanation was that of Wood (1950). He considered
those systems whose relative radii had been reasonably well determined
from reliable photometric observations and for which a value of the
mass ratio could be found. Of these, 19 had shown irregular period
fluctuations. For these systems it was possible to compare the radii
of the components obtained from the light curves with the sizes of
the Jacobian or Roche lobes computed from the mass ratios. With one
exception, every system which had shown erratic period changes had one

component in contact with this lobe. The exception, AR Lacertae, was known even then to be unusual for reasons other than period changes - suspected intrinsic variability, for example, and the intermittent occurence of emission lines in the spectrum. In every case, except this one, the components approached instability at the end of the shorter equatorial axis, not at the inner Lagrangian point, indicating on the evidence then available that the real stars were not as distorted as the limiting surfaces which are computed with a number of simplifying assumptions. It would be interesting to repeat such a study with modern methods of solution and the far greater number of well observed light curves now available, but the difficulties of trying to obtain the shape of a star from the light curve are well known, and if, as seems to be necessary the shapes are postulated first, the test carries low weight.

Not only did the systems showing sudden period changes have one component touching the limiting surface, but no such changes were found in well observed systems which had neither component near its surface. The suggestion was made that close double star systems could be divided into two classes using this criterion. Wood's classification was later adopted by Kopal, who suggested the nomenclature of "detached" and "semi-detached" now currently in use.

A physical mechanism to explain the changes was also considered in the 1950 paper. If the ejection is by violent prominence activity somewhat similar to that on the sun, the change of period will depend not only on the mass loss, but also on the place and velocity of ejection. Simplified calculations, assuming ejection at either end of the shorter equatorial axis where the systems seemed nearest to instability, indicated the changes by ejections at reasonable velocity would be much larger than changes caused by mass loss alone and that involved for some of the larger observed changes was computed to be 10^{-7} $\mathfrak{M}_{\odot}$ /year. This seemed large at the time, but is very much smaller than rates now assumed in current evolutionary theories. It should be mentioned that from present observational data it is extremely difficult to distinguish between one relatively large period change and a series of smaller ones having the same cumulative effect.

Other, more sophisticated, studies of mass loss followed this original, rather primitive, suggestion. A study of Huang (1963) derived formulae showing increase and decrease of period. Piotrowski (1964) and Kruszewski (1964) made thorough investigations which give detailed formulae for circular orbits. Kwee (1955) computed period changes in relation to total angular momentum and masses.

Later work on mass loss and period changes especially in contact binaries, has been carried out by many authors (e.g., Van't Veer, 1972) who have considered loss of angular momentum, and the effect on the orbital elements caused by mass ejection. Two recent papers in particular deserve mention. Hall (1972) has postulated a region of tremendous starspot activity on one side of the cooler component of RS Canum Venaticorum ejections from which cause the observed changes in orbital period, and Herczeg and Frieboes-Conde (1974) in studying RZ Cassiopeiae find 10 sudden period changes within a 32 year interval.

Period variations can also be caused by mass transfer within the system itself. In discussing period variations in eruptive binaries, Smak (1972), noted that in the three best studied cases, the changes are of an alternating character. He wrote an equation describing

the changes in which the first term was due to changes in orbital
motion, the second to mass loss from the secondary component and the
third to mass loss from the system; this third term he felt could
safely be neglected. Variations were explained by exchanges between
the orbital momentum and the rotational moment of the components and
of the disk surrounding the primary; this, in turn can either be due
to the transfer of mass carrying an excess of momentum, positive or
negative, or due to the tidal interaction. The picture can be simplified by assuming that an exchange between orbital momentum and the
rotational momentum of the disk is much more important than exchange
with the rotational moments of the components.

Biermann and Hall (1973) used these ideas in developing a model
to explain alternate period changes in "Algol-like" binaries. Their
hypothesis suggests that sudden bursts of mass are transferred to
the hot star, possibly as a result of an instability of the type suggested by Bath (1972). The mass and the angular momentum thus lost
from the cooler star is stored temporarily as rotation in the surface
layers of the hotter component, or onto a disk, thereby decreasing
the orbital period, and then goes back into orbit as increased turbulence "lets the friction time scale catch up with the dynamical
time scale of the mass transfer (thereby increasing the period)".
Both mechanisms call for a sudden release of the energy of the ionization layers of hydrogen and helium. This has been criticized by
Plavec et al. (1973) on various grounds; however, a sudden transfer
of mass from any cause would satisfy the requirements of the Biermann
-Hall hypothesis. Hall (1975) later developed this model more fully
and summarized the overall problem of period changes caused by mass
loss or mass transfer.

An unusual example of mass transfer in the case of SV Centauri has
been discussed by Wilson and Starr (1976) in their analysis of observations published by Irwin and Landolt. This system is found to be
an early type contact binary and the degree of "overcontact" is large
($\sim$ 0.7). It shows a very large value of dP (period change) - indeed
one of the largest known in terms of dP/P. These authors suggest,
for lack of a realistic alternative, that this change is almost certainly due to mass transfer, with the flow from the more to the less
massive star, with conservation of mass and angular momentum. The
negative value of dP/dt occurs when the flow is from the more to the
less massive component. Very few, if any, other systems have been
observed in this stage of rapid mass transfer because it lasts for so
short a time in the evolutionary history of the system. From the
period changes, Wilson and Starr compute the present mass transfer
rate to be 4 x 10^{-4} $\mathfrak{M}_\odot$/year or about 30 times that for β Lyrae.

We emphasize that the large period changes found in the RS Canum
Venaticorum variables are still not well understood. These are all
detached systems and the energy required to eject large amounts of
mass must be very large indeed.

5. SUMMARY

While the preceding discussion by no means covers all the work done
or papers published especially in recent years, it does give some idea
of the importance of period studies. This is in sharp contrast to the
attitude of thirty or fourty years ago; the only purpose of a period

study, in the eyes of many astronomers, was to derive light elements
of maximum precision to aid in planning observing programs. The
importance of continuing observations is difficult to overemphasize.
This is an outstanding example of the value of observations increasing
with age. Even two or three rather crudely determined minima of sixty
or seventy years ago can be essential in period studies; by the same
token, minima observed today will be of increasing future value.

Since there are not enough professional astronomers to keep all
the eclipsing systems under proper observation, this is a field where
work of amateurs is of value. At least three such groups - one section
of the (AAVSO) American Association of Variable Stars Observers, (BAV)
Berliner Arbeitgemeinschaft f. Veranderliche Sterne, and the (BBSAG)
Bedeckungsveranderlichen Beobachter der Schweizerischen Astronomischen
Gesellschaft are carrying on sustained programs, and others will be
welcome.

CHAPTER 7

EVOLUTION OF CLOSE BINARIES

1. COMPUTATION OF PARTICLE TRAJECTORIES

When the massive spectrographic investigation of close binary systems that was undertaken during the decades of the 40's and 50's made it evident that the presence of a gaseous structure was a common phenomenon, it was not clear whether the streams and the rings were of any significance from the point of view of the evolution of the objects concerned. Even in 1960 Struve (1962) stated that "there were times in my own experience when I thought that perhaps the existence of gas streams would have a very significant effect upon the evolutionary track of a star. Other times I felt that this phenomenon were superficial and they did not seriously modify the evolutionary track. And I think there is no consistent answer". It was only six years later that the first computations of close binary evolution were reported and a new epoch in the history of the field began.

An attempt to put together the observational data and try to draw an evolutionary picture was made by Struve (1950a) in a book entitled "Stellar Evolution, an Exploration from the Observatory" which collected the Vanexum Lectures he delivered at Princeton in 1949 when our ideas on the evolution of single stars were still in their infancy. In his conclusions Struve was led to discuss the possible past and future history of a W Ursae Majoris system: out of an interstellar cloud an early type single star with large amount of angular momentum would be formed; an explosive process would convert the single star into a U Coronae Borealis-like binary, this, through mass loss, would turn into a W Ursae Majoris system, the latter becoming a planetary system in a sort of final stage.

When large electronic computers became available the time was still not ripe to tackle quantitatively the problem of the evolution of a close binary system but it was possible to take the important step of computing particle trajectories to find out whether the gaseous rings that exist around the primary component in systems such as RW Tauri could actually be produced by the matter that flows in the gaseous stream from the secondary. The first computations were made by Kopal (1956, 1957) and by Gould (1957) by considering only gravitational forces and by assuming that the particles originate in the component with smaller mass. The modes of ejection were different in the views of the two authors and several mass-ratios and velocities were considered. It was clear from this work that in some cases a gaseous ring ought to be formed around the star towards which the stream moves. Gould (1959) considered also the case of the ejection coming from the more massive component.

Computations of this kind had been started by Günther (1955) without the benefits of large computers and the case of SX Cassiopeiae was specifically considered by him. By also applying numerical integration of the equations of the restricted three-body problem Plavec et al. (1964; see also Plavec, 1966) tried to consider the cases that correspond to specific systems and determine for which values of the parameters a gaseous ring is likely to be formed. The result, in agreement with what Mrs. Gould had found, is that the ring-forming orbits require relatively large initial velocities.

Since the gaseous streams are dense enough to produce absorption features, Struve (1957) pointed out earlier the importance of considering the problem of the trajectories from the hydrodynamical point of view. This was done first by Pendergast (1960) and more recently by Biermann (1971) who assumed that the stream is a supersonic flow. The tendency to form a ring around the component towards which the stream goes was found "for all values of the parameters used in the calculations". More recently, Pendergast and Taam (1974) and Flannery (1975) developed hydrodynamical treatments of the gas flow which permit taking into account pressure terms, shock formation, etc. and they applied these to cases resembling those of U Cephei and Z Camelopardalis, respectively.

Another attempt to try to understand the evolution of close binaries was through the classification scheme suggested by Sahade (1960a, 1960b).

2.   CLOSE BINARY EVOLUTION

Early on it was clear that the Algol-type systems presented a problem, the so-called "Algol paradox", in the sense of the mass-luminosity relation; and the paradox was more profound, so to say, when we began to understand the evolutionary tracks of single stars. Although it is well known what is meant by the Algol paradox, and we have mentioned it several times in Chapter 1, it may be well, for the sake of completeness, to repeat here that the Algol-type systems are formed by a main sequence A-F type primary and a subgiant G-K secondary, the latter being less massive than the former, although its position in the Hertzsprung-Russell diagram would suggest that this star should have evolved faster.

Crawford (1955; cf. Struve, 1955) and Kopal (1955) were the first to postulate independently that the paradox should have resulted from the fact that the present secondaries were initially the more massive components, and that, in the course of their evolution they expanded and that after their dimensions were such that they filled the inner zero-velocity surfaces any further expansion must have signified a loss of mass. This mass was either collected by the companion or ejected from the system thus finally reversing the situation as far as the masses are concerned. This very attractive explanation was further elaborated by Morton (1960) who utilized the knowledge of the time on the evolutionary tracks towards the red giant stage. Morton, who assumed constancy of period throughout, was able to show that the

transition between the stage where the original primary is still the
more massive component and the stage in which this star becomes the
less massive component of the pair occurs in a Kelvin or thermal time
scale. Smak (1962) tried to extend the discussion with the utilization
of several approximate models with helium burning in the core and
hydrogen burning in a shell and suggested that the highly overluminous
subgiants are objects in the helium burning stage of their evolution.

Only when more detailed evolutionary tracks became available, particularly those of Iben (1965a,b; 1966a,b,c; 1969a,b) the time was ripe
for taking up the problem of the evolution of a close binary system
in a more systematic and thorough manner. The colloquium on the evolution of double stars held in Uccle in 1966 was the occasion to learn
of the steps that had already been taken in that direction at three
different centers, namely, in Göttingen by Kippenhahn and his group,
in Warsaw by Paczyński and Ziólkowski and in Ondrejov by Plavec and
his coworkers. The problem was also going to be undertaken by Wrubel
at Indiana but the idea was abandoned when he found that the German
group was already well ahead.

Several review papers have been published - particularly by Plavec
(1968, 1970, 1973) and by Paczyński (1970a, 1971) which give an account
of the state of the art at the time each one was written and describe
in a very comprehensive and clear way our understanding of the matter.
More recent review papers by Thomas (1976) and by Flannery (1976) deal
with consequences and effects of mass transfer in close pairs.

In our era of computations of close binary evolution, a little over
a decade old, we can already talk about three different epochs. In
the beginning the queries were naturally concentrated in the mass-
losing star and only conservative cases were considered, a few years
ago non-conservative cases began to be taken into account and, then
even more recently, attention is being directed also to the evolution
of the mass-gaining star.

Of course the computations are carried out under a number of simpli-
fing assumptions which can be listed as follows:

   a) the star is spherically symmetric, that is, rotational
       and tidal distortions are neglected;

   b) the star is in hydrostatic equilibrium;

   c) the orbit is circular;

   d) mass transfer takes place when the stellar radius
       exceeds a critical radius which is the radius of the
       Roche lobe;

   e) the stellar radius is always equal or smaller than the
       Roche lobe;

   f) there is a conservation of the total mass and the
       orbital angular momentum, thus implying that the whole
       of the matter being lost by one component is being
       accreted by the other one.

Condition (f) defines, of course, the conservative models.

In his 1971 review paper, Paczyński analyzed these assumptions
and stated that condition e), "cannot be satisfied if the primary component has a deep convection envelope when mass loss commences"
(Paczyński, 1965; Paczyński et al., 1969; Refsdal and Weigert, 1967;
Lauterborn, 1970; Lauterborn and Weigert, 1972) and it becomes impossible

"when the two Roche lobes are filled up by the two components which
display a tendency toward expanding even more". In regard to the
condition of hydrostatic equilibrium Paczyński adds that it "is violated in a small region close to the stellar surface and may be impro-
per for primaries with deep convective envelopes, for secondaries
while the rate of mass accretion is very high, and for contact systems
with components that tend to expand beyond their Roche lobes".

The condition that deserves special thought is the one that postu-
lates conservation of mass and of angular momentum. Since the stage
of drastic mass loss that leads to the reversal of the original mass
situation in the system is very rapid, the mass loss should be charac-
terised by large velocities or large densities of ejection and, therefore, simple reasoning would suggest that condition f) is in all like-
lihood not usable. Moreover, Plavec (1973) in his 1972 invited paper
at the I.A.U. Symposium No. 51, stressed the fact that a number of
Algol systems cannot be explained in their present evolutionary stage
with the present total mass without accepting the fact that a certain
amount of mass should have been lost to the system.

Guseinov and Novruzova (1976) have also analyzed the problem in the
light of the fact that pairs develop into systems with a white dwarf
component and concluded that in the stage of rapid mass loss a very
large part, and sometimes practically all, of the mass shed by the
originally primary is lost to the system.

Paczyński and Ziółkowski (1967) were first in formulating a non-
conservative model and they did this by introducing two parameters,
which represented the fraction of mass lost to the system and the rate
of loss of orbital angular momentum, respectively. In his review paper
Plavec mentions an unpublished formulation of the non-conservative case
made by Weigert and Lauterborn in 1970. In this model matter is supposed to leave the system through the Lagrangian point L_2.

Since practically all systems are observed in the stage in which
mass loss proceeds at a slow pace, that is, in a stage in which evolution proceeds on a nuclear time scale, the observation of actual systems does not provide information about the amount, the rate, or the
fate of mass loss in the stage that proceeds on a thermal time scale.
Thus we need to find some way of approaching the problem of the physics
of mass loss throughout the evolution of the close pair.

3.   COMPUTED CASES

As we have mentioned in Chapter 1, four models or cases could result
in close binary star evolution, namely,

Case A, when the initially more massive component fills the
lobe that contains L_1 while it still burns H in its core,

Case B, when the filling of the critical lobe occurs while
the nuclear energy generation results from H-burning in a shell that
surrounds the H-depleted core, that is, while the star is moving
across the Hertzsprung gap in the way to become a giant;

Case C, when the outflow of the Roche lobe corresponds to
a phase while the energy generation results from He-burning and
H-burning shells;

Case AB, when the lobe filling stage is reached in case A

and the outflow proceeds up to the situation depicted in case B.

The nomenclature which is now commonly used was first suggested by Kippenhahn and Weigert (1967) for the first two modes, while case AB was discovered by Ziółkowski (1969a, 1969b), and a case C was first computed by Lauterborn (1969). For a system with components of 5 and 2.5 solar masses, case A occurs for periods between 0.65 and 1.5 days, case B for periods between 1.5 and 87 days, and case C for periods between 87 and 4300 days (Paczyński, 1970b).

Paczyński (1971) lists the series of papers produced through 1970 in which the evolution of close binaries in different combinations of initial masses and in different modes was computed.

From the start one of the important aims of the computations was to try to produce a white dwarf as the result of the evolution of the mass-losing component of a close binary. Kippenhahn et al. (1967) first showed that such a result is obtained if one considers mode B for small masses (cf. also Weigert, 1968). The pair that was considered was formed initially by two main sequence stars of 1 and 2 $\mathfrak{M}_\odot$, respectively, separated by 6.6 $R_\odot$. Other cases were computed later (eight in total) and a generalization of the different results obtained was offered by Refsdal and Weigert (1971). The initial systems had total masses of 2.5, 3.0 and 3.5 $\mathfrak{M}_\odot$ and separations that ranged from 6.0 to 145 $R_\odot$, but the final masses of the original primary component did not differ very much from one another, the actual range being from 0.43 $\mathfrak{M}_\odot$ for a total mass of 2.5 $\mathfrak{M}_\odot$ and a separation of 145 $R_\odot$, to 0.26 $\mathfrak{M}_\odot$ for a total mass of 3.0 $\mathfrak{M}_\odot$ and a separation of 6.6 $R_\odot$.

The interest in producing a close binary with a white dwarf component was accompanied by a desire to consider systems which would turn into a Wolf-Rayet binary. It was Paczyński (1967a, 1967b) who first suggested that case B of binary evolution of a massive pair with the original mass between 5 and 65 $\mathfrak{M}_\odot$ is likely to produce a Wolf-Rayet binary. It is actually believed that the originally primary component should then settle down near the main sequence for the helium burning stars, a stage which is thought to correspond to that of the Wolf-Rayet stars. We shall refer to this point in Chapter 9.

According to van den Heuvel (1973, 1976) Wolf-Rayet binaries are the progenitors of the X-ray doubles. Naturally the emphasis has been largely placed in establishing the mass configuration and the location of the components in the Hertzsprung-Russell diagram, but this is certainly not enough to identify the end-product of binary evolution with particular types of objects. Underhill (1969) pointed out in a discussion related to the binary evolution that presumably would lead to a Wolf-Rayet binary. "If we are to identify models in a certain part of the H-R diagram with Wolf-Rayet stars, it is necessary to show that the typical type of spectrum will certainly occur and that the postulated theoretical models give not only necessary but also sufficient physical conditions for generating the spectrum." And we could certainly generalize this statement (Sahade, 1968).

Regarding the evolution of the mass-gaining component we would at this time only say that the computations made so far by Benson (1970) and by Ulrich and Burger (1976) suggest that the star could become oversized and overluminous depending on the rate of accretion.

4. EFFECTS OF CLOSE BINARY EVOLUTION

4.1   Because of the mass outflow from one component, the mass gaining
by the companion and the mass loss to the system, the evolution of
interacting binaries follows a track that departs considerably from
that of the evolution of single stars. Moreover, the physical characteristics of the elements of the gaseous structure, which could
vary widely from one object to the other, and must be related to the
type of components involved and to the parameters of the mass outflow,
could not only produce very strange stars, from the point of view of
their spectra and even of their luminosity behavior, but also a masking
and/or distortion of the stellar features such that the binary nature
of the object is very difficult to disclose.

As to the strange stars that result from binary evolution, we could
mention the Algol systems, β Lyrae, the symbiotic objects, the cataclysmic variables, etc. A very good example of a system where pure
stellar features are in fact absent and the light curve is quite distorted, is provided by W Serpentis, a system to which we have referred in Chapter 5. As we have also mentioned in Chapter 5 (7.2), a
careful analysis of the observations has led Peters and Polidan to the
interpretation of HR 2142 (cf. Peters, 1973), a Be star which displays a periodic shell phase, as a 80 day-period binary system undergoing mass outflow. Let us add the case of the infrared source HD
101584 which is explained as a single-lined binary with an unseen cool
companion that is losing mass (Humphreys, 1976).

4.2   A binary component which has experienced a large amount of mass
loss in its evolution in a mode other than A, should show an abnormal
chemical abundance if material from the deep interior is brought out
to the surface. But in order to be detectable such departures from
normal composition should be substantial. In particular, if evolution
has proceeded in the context of mode B until our star has become a
helium star, we should certainly observe that the particular component
has a He-enriched atmosphere.

Three binaries have been reported to abide to such an expectation,
namely, β Lyrae (Boyarchuck, 1959; Struve and Zebergs, 1961; Hack and
Job, 1965), υ Sagittarii (Greenstein, 1940) and KS Persei (=HD 30353)
(Nariai, 1963; Wallerstein, Greene and Tomley, 1967). We wonder how
much meaning one can assign to a curve-of-growth analysis of a peculiar
spectrum which is as full of emission lines as in the case of β Lyrae,
or whether we could then have strong H emission and an outer envelope
that displays lines of H and of HeI, as it is the case in υ Sagittarii,

Several years ago Hall (1967) made a narrow band photometric study
of a number of Algol subgiant secondaries and found that possibly CN
is deficient by a factor of 3 or so, a marginal result which was confirmed by Spinrad's scanner observations of AR Lacertae, "a weak-line
star, not only in the CN-band, but also in metallic lines in the blue
when it is in total eclipse" (Popper, 1970). Similarly, the Algol
type southern binary S Velorum was investigated by Sisteró (1968, 1971).
Sisteró found an ultraviolet excess which could be interpreted in terms
of low metal abundance, a result which was confirmed by Bond (1972) who
reached the conclusion that the secondary component of S Velorum is
metal-deficient. However, during the 1974 outburst detected by Batten
et al. (1975) in U Cephei, it was found that the ultraviolet excess increased by up to 0.2 magnitude over the "normal" value of 0.1 magnitude.
These observations throw strong doubts to an interpretation of the ultraviolet excess in Algol secondaries entirely in terms of metal-deficiency.

As possible indication of abundance anomalies we could also mention the difficulties in classifying the spectrum of systems like U Cephei (Batten, 1970).

We should further mention Kondo's (1968) investigation of possible abundance anomalies in close binaries of spectral types A0-A2 and F5-F6. This work was carried as a follow-up of Abt's (1965) suggestion that all close binaries with periods of about 1-100 days and with spectral types A4-F2 (IV-V) are Am stars. Kondo's conclusion was that perhaps "metallicity is a norm rather than a peculiarity for close binaries, at least within a certain spectral range".

In a recent paper Plavec and Polidan (1976) mention a number of investigations which suggest metal-deficiency in secondaries of Algol-type systems and in two RS Canum Venaticorum binaries. But again the results are in the limit of significance and the methods used cannot overcome the uncertainties that result from the type of objects involved and the observational limitations imposed by the equipment or the faintness of the system under consideration.

5. PRE-MAIN SEQUENCE STAGE

No significant computations have been carried out that would try to cover binary evolution in the stage prior to the main sequence, nor do we know exactly how a binary system is born although the fission theory is becoming more and more attractive to explain systems like the contact binaries.

Likewise, observations have not provided us with cases that could unmistakably be understood in terms of contraction toward the main sequence. The primary of the system KO Aquilae was for some time believed to be one such case, because the secondary component was supposed to be an undersize subgiant - that is, an object typical of a group that was later shown to be non-existant. At the present time the best candidates of being binaries with one component still in the contracting stage are BM Orionis (Hall, 1971; Popper and Plavec, 1974, 1976), as one possible alternative interpretation, and the RS Canum Venaticorum binaries, the latter group being discussed more in detail in Chapter 9.

As far as BM Orionis, an eclipsing star which is the faintest of the four stars in the Orion Trapezium, the secondary component appears to be highly flattened in the approximate shape that is predicted by theory for a pre-main-sequence star undergoing rapid rotation (Boden-heimer and Ostriker, 1970).

Actually, as McCluskey et al. (1975) have put it, "it is not an exaggeration to state that our understanding of the pre-main-sequence evolution of close binaries is poor".

6. AXIAL ROTATION

One problem that we should mention is the context of binary evolution is the lack of synchronization between the rotational and the

orbital velocity that characterizes the primary component of a number
of Algol-type binaries. A typical case is that of U Cephei where the
velocity of rotation of the early type main sequence primary is nearly
four times the velocity in its orbit. (see Fig. 5.1, Chapter 5).
The list includes RY Persei, S Velorum, RZ Scuti, U Sagittae, SW
Cygni, AQ Pegasi, AR Cassiopeiae, and as Struve (1950b) already pointed
out, "in systems with $\alpha = \mathfrak{M}_2/\mathfrak{M}_1 = 1$, there is almost never such a difference in rotational period".

The meaning of the departure from snychronization is still not
understood. Batten (1975) in his thorough investigation of U Cephei
suggests that the "rapid rotation is a surface phenomenon, and the
underlying star may not be rotating so rapidly. The outer layers of
the star may be carried around by the 'disk', the inner portions of
which, as we have seen, may be rotating at more than 600 km sec^{-1}."

The amount of information relevant to the understanding of close
binary evolution has increased tremendously since the times of the
strong observational effort that was started at the Yerkes and McDonald
Observatories in the 1940's. The range of objects that are now known
as members of close binary systems include stars that are in the very
late stages of their lives like the undermassive component in HZ 22
(Greenstein, 1973). It also includes compact objects like the white
dwarf in V471 Tauri (Nelson and Young, 1976), the pulsar PSR 1913+16
(Taylor and Hulse, 1974) and the X-ray sources in X-ray binaries. We
have learned that the central stars of planetary nebulae are in several
cases close doubles. And as we said earlier in the Chapter, careful
analysis of many bizarre objects are leading to the conclusion that
they are interacting physical pairs. But we still lack a unified picture. However, the possibilities for gathering data in a wide energy
interval, with increasing sensitivity, and for learning about a more
detailed behavior of the systems in different parts of the volume they
occupy, will certainly lead to dramatic progress in the field.

CHAPTER 8

SPACE ASTRONOMY

1. INTRODUCTION

The limitations and drawbacks for astronomical research that result from the presence of the Earth's atmosphere made astronomers
eager to try to find ways to overcome this barrier. The first UV
spectrum of an astronomical source was taken immediately after the
Second World War, in 1946, by means of a V-2 rocket launched in the
United States, but it was only about two decades later that space
astronomy emerged as a new and vigorous branch of our science.

The possibilities opened up by the use of space vehicles and bal-
loons have been - and will continue to be - fully utilized for the
investigation of close binaries to the extent of the limitations set
by the detecting equipment. Thus we have been able to extend our
knowledge of the behavior of a number of systems to ranges of energies other than those observable with ground-based telescopes.
Moreover, the discovery of unexpected types of systems has made
close binary stars one of the most lively, exciting and flourishing
fields of research in Astronomy today.

The first observations of close pairs, both in the UV and in X-
radiation, were made with rockets. In fact, the first non-solar X-
ray source to be discovered was found to be a close binary more than
a decade after Giacconi detected it in 1962. This source is Scorpius X-1, the strongest one in the sky in the 1-10 keV range, and
its discovery, as has been true with many significant discoveries,
was merely accidental and resulted from a failure to reach the plan-
ned objective of the flight.

2. UV OBSERVATIONS

Ultraviolet observations are extremely important in deriving information about the outermost layers of the gaseous envelopes and
in ascertaining the presence of hot companions in close binary systems.

The first close pair observed in the ultraviolet was the Wolf-
Rayet binary γ_2 Velorum (Carruthers, 1968). The stellar flux was
found to be deficient below Ly-α, a feature that is common to

ζ Puppis, and was traced to absorption by the extended atmospheres or gaseous envelopes that surround these objects. Moreover, as mentioned in Chapter 5, Stecher (1968) discovered in the spectrum of γ_2 Velorum the presence of the intercombination line C III] λ 1909 which arises in a relatively tenuous mass of gas that is moving away from the system at a velocity of the order of 1500 km sec^{-1} (see also Morton et al., 1969). This velocity agrees fairly well with the velocity derived from the diluted lines of He I in the photographic and near infrared regions, as also mentioned in Chapter 5. The ultraviolet spectrum of γ_2 Velorum is rich in P Cygni-type profiles which seem to originate in layers that have even larger velocities of approach (Johnson, 1977). Therefore, we are now in possession of information that refers to different layers of the outer envelope of the system.

The OAO-2 satellite provided the first opportunity to obtain ultraviolet light curves of variable stars and β Lyrae was a close double on the program. Light curves of β Lyrae were obtained in six different wavelengths, namely, λλ 1380, 1500, 1920, 2460, 2980, and 3300 Å, with filters that had a typical half-width of 200 Å (Kondo et al., 1971). The results which were expected to throw some light on the nature of the secondary component of the system, can be sum-marized as follows:

a) the eclipses as well as the maxima are asymmetric in all wavelengths;

b) the minima are deep, except for λ 1920 Å; at this wavelength they are shallower;

c) in the shortest wavelengths, except for λ 1920 Å, the second-ary minima appear to be about as deep as the primaries.

This behavior, at least for the three shortest wavelengths, can be understood in terms of the coincidence of the wavelength regions with strong emission features and of the location in the system of the gaseous masses where such emissions originate (Hack, 1974; Hack et al., 1976a). No new infomation resulted about the companion to the B8 II component.

Observations of close binaries in the ultraviolet have been also obtained with spectrometers flown aboard the TD1-A and the OAO-3 (Copernicus) satellites, with a low resolution experiment from the Apollo 17 (α Virginis), and with an objective prism spectrograph from the Skylab. We shall leave for Chapter 9 the results regarding the Wolf-Rayet binaries and for Chapter 10 those related to β Persei and to β Lyrae. Let us only mention now two further interesting results.

One of such results refer to the Of binary 29 UW Canis Majoris. The material collected with the Copernicus satellite in the wave-length interval λλ 950-1560 Å provided information about the gaseous envelope in the system through the display of a number of stellar P Cygni-type profiles of CIII, CIV, NV, SiIV, SiV, PV, FeIII. The absorptions yielded radial velocities that are in the range of -200 to -800 km sec^{-1} and led to an average value for the mass loss rate of $3 \times 10^{-6} \, \mathfrak{M}_\odot$ yr^{-1} (McCluskey et al., 1975).

The second point refers to the discovery of a hot companion in HR 3080 = a Puppis, previously known as a single-lined spectro-scopic binary of spectral type G5III and a period of 2660 days.

The far ultraviolet spectrum of the object - extending to about 1700 Å - observed with the Skylab objective prism spectrograph, disclosed the nature of the companion (Parsons et al., 1976), thus making the binary a most likely member of the group of the so-called symbiotic stars to which we shall refer to in Chapter 9.

3.   X-RAY BINARIES

The first satellite designed for X-ray non-solar astronomy was the SAS-1, better known as the Uhuru satellite, launched in December, 1970, with its sensitivity in the range of 2-6 keV.   The information provided by the Uhuru was extremely rich and opened up new avenues of research in galactic and extragalactic astronomy.   The discovery, from the Uhuru data, of the existence of close binary systems with X-ray source components meant a breakthrough in our knowledge of the very late stages of their evolution.

The sky survey carried out by Uhuru has led to a catalogue of 161 X-ray sources, designated as 3U (for Third Uhuru Catalogue) sources, a number of which have already been identified with known stars. Those that have been found to be close binaries are listed in Table 8.1.

The systems listed can be divided into two different groups that we have designated as A and B, respectively.   The systems in group A are characterized by the fact that the companions to the X-ray component are O and B giants or supergiants.   On the other hand, group B deals with short period systems where the companions to the X-ray objects are much smaller and less luminous than those in group A. Some of them are perhaps late type subgiants, as is suggested by Cygnus X-2.   We have not included in Table 8.1 the X-ray source Circinus X-1 (=3U 1516 -56), which has been identified optically with a faint red emission object, because of the lack of sufficient information.

4.   THE NATURE OF THE X-RAY BINARIES

In Chapter 5 we have described our present understanding of the origin of the X-radiation in X-ray binaries, in terms of the thermal energy released when the particles that stream from one of the components are accreted by the highly compact companion.   We are thus dealing with the conversion of the particles' kinetic energy, which is proportional to its radius, into thermal energy.   Early on Hayakawa and Matsuoko (1964) and Zeldovich and Guseynev (1966) suggested that the infall of matter onto compact objects could release energy detectable in the X-ray range.   Moreover, the optical identification of Scorpius X-1 with a faint blue object (Sandage et al., 1966), the spectrum of which resembled that of an old nova, prompted several theoreticians (Novikov and Zeldovich, 1964; Burbidge, 1969; Shklovsky, 1967) to suggest that X-ray sources were actually interacting binary systems with a neutron star, or perhaps a black hole as one of the components.

The observed X-ray luminosities, which are in the range of 10^{36}-

Table 8.1

X-RAY BINARIES

Designation of X-ray source	Binary Period (days)	Optical Companion			Remarks
		Spectral type	Magnitude	Designation	
3U 0115 -37 SMC X-1	3.9	B0 I	13	Sk 160	$0^s.7$ pulse period X-ray eclipses
A 0535 +26		B0 IIIe	9	HDE 245770	104^s pulse period
3U 0900-40 Vela XR-1	8.9	B0.5 Iab	6	HD 77581	283^s pulse period X-ray eclipses
1118 -60 CentaurusX-3	2.1	O6.5 III	13	V799 Centauri (Krzemiński's star)	4.8 pulse period X-ray eclipses
1700 -37	3.4	O6f	6	HD 153919	X-ray eclipses
3U 1956 +35 Cygnus X-1	5.6	O9.7 Iab	9	HDE 226868	radio emission
3U 1617 -15 ScorpiusX-1	0.8	nova-like	12.4-13.6	V818 Scorpii	
1653 +35 Hercules X-1	1.7	late A	13.1-14.7	HZ Herculis	$1^s.2$ pulse period X-ray eclipses; radio emission
1809 +50	0.13	CV-like		AM Herculis	
2030 +40 Cygnus X-3	0.2				X-ray eclipses; radio emission
3U 2142 +38 Cygnus X-2		sdG	14		radio emission

A { (first seven rows)

B { (last four rows)

10^{38} erg sec^{-1}, and the temperatures of 10^8 °K which are required if
the radiation mechanism is to be thermal, indeed imply that the accreting object must be a <u>collapsed</u> one.

It is at present generally believed that such an explanation must
hold true for all galactic X-ray sources that are not associated with
supernova remnants. Furthermore it is admitted that the matter that,
through accretion, should give rise to the X-ray emission is supplied
by the companion to the X-ray source through a radiation pressure-
driven stellar wind or through the "classical" mechanism of inner
contact surface overflow. The latter process appears to be at work
in Hercules X-1 and, therefore, it must probably be at work in all
the systems of group B. The stellar wind mechanism should, on the
other hand, operate in our group A of X-ray binaries. A disk must
form around the accreting star, because of the large angular momentum of the accreted material (Prendergast and Burbidge, 1968), and
therefore, actual accretion requires that dissipation of the angular
momentum takes place. Recently Ostriker <u>et al.</u> (1976) have considered
the effect of pre-heating of the infalling gas by the emergent X-radiation and concluded that preheating can, in general, suppress the
accretion. The exceptions correspond to the cases where shadow effects
are present, and to the cases where the stellar wind is the supply
mechanism, provided the relative velocity between the gas and the
accreting star is supersonic.

5.   PULSATING X-RAY BINARIES

The hypothesis that X-ray sources like Scorpius X-1 could be close
binary systems with one compact component gained momentum when, in
1971, the <u>Uhuru</u> observations revealed two pulsating X-ray objects that
apparently underwent eclipses. These two sources were Centaurus X-3
and Hercules X-1, which pulsate with periods of 4.8 and 1.2 seconds,
respectively. In both objects two distinct levels of intensity were
observed, which repeated with a period of nearly 2.1 days, in the
case of Centaurus X-3, and of 1.7 days, in the case of Hercules X-1,
and were interpreted as indicating eclipses and, therefore, the existence of binary motion. In support of this interpretation it was
further found that there is a periodic variation of the pulsation
periods that suggested a Doppler effect from the orbital motion; the
pulsation periods increased as the X-ray sources came out of eclipse
and decreased when the sources were supposedly moving away. Table
8.1 list the known pulsation periods of the X-ray binaries as compiled
by Rappaport and Joss (1977a), who also include in their discussion
of binary X-ray pulsars four additional objects which are suspected
of being binaries. With the addition of A 1540 -53 they are listed
in Table 8.2 together with the pulsation periods and their "optical"
counterparts.

The pulsation periods that we have quoted suggest that the accreting compact objects we are dealing with are white dwarfs or, as it
is more generally believed, rotating magnetized neutron stars (cf.
Joss and Rappaport, 1976). Rappaport and Joss (1977b) have compared,
in the case of seven X-ray pulsars, the secular changes that would
result from the torques exerted as matter is accreted onto the compact
object. The computations were made on the assumptions that "(a) the
pulse period reflects the rotation period of a compact objects; (b)

Table 8.2

PULSATING X-RAY SOURCES SUSPECTED OF BEING BINARIES

X-ray Source	Pulsation Period (seconds)	Optical identification	Notes
3U 0352 +30	835	X Persei	1,2
A 1118 -61	405	RS Centauri	2,3,4
3U 1223 -62 ? GX 301-2	696	WRA 977 (B1.5 Ia)	5,6
A 1540 -53	529		7
3U 1728 -24 GX 1 + 4	122	gM (symbiotic or recurrent nova)	8

Notes to Table 8.2:

1. cf. Margon, B., X-Ray Binaries (NASA SP -389), ed. E. Boldt and Y. Kondo, p. 719 (1976).
2. Canizares, C. R., Backman, D. E., Jernigan, J. G., McClintock, J. E. and Nugent, J. I., preprint (1977).
3. Chevalier, C. and Ilovaisky, S., IAU Circ. No. 2778 (1975).
4. Fabian, A. C., Pringle, J. E., and Webbink, R. F., Astrophys. Space Sc. 42, 161 (1976).
5. Vidal, N. V., Astrophys. J. Letters 186, L81 (1973).
6. Jones, C. A., Chetin, T. and Liller, W., Astrophys. J. Letters 190, L1 (1974).
7. Becker, R. H., Swank, J. H., Boldt, E. A., Holt, S. S., Pravdo, S. H., Saba, J. R. and Serlemitsos, P. J. Astrophys. J. Letters 216, L11 (1977).
8. cf. Davidsen, A., Malina, R. and Bewyer, S., X-Ray Binaries (NASA SP -389), ed. E. Boldt and Y. Kondo, p. 691 (1976).

the accretion is mediated by a disk surrounding the compact object
and rotating in the same sense, and (c) the compact object is a neutron star". The agreement obtained was fairly good lending thus support to the picture that corresponds to the assumptions.

6.   OTHER POSSIBLE X-RAY BINARIES

As time goes by and we acquire more information about the behavior
of the X-ray sources and optical identifications become available,
we find further evidence in favor of the hypothesis of their largely
being close binary systems. The pulsating and the transient, and
even the variable, X-ray sources are good candidates. Two X-ray novae
should be mentioned in this context. One is Triangulum Asutrale X-1,
the position of which appears to coincide with that of an optical
novae (cf. Bradt et al., 1977a). The other one is A 06200, the optical counterpart of which is V616 Monocerotis, also a nova (cf. Matilsky
et al.,1976). In the latter case, both the X-ray and the optical observations appear to be represented by a period of 7.4 days (Tsumeni
et al., 1977)

Recently two X-ray sources have been identified with Be stars.
These sources are 1S 0053+64 which appears to coincide in position
with the well known B0.5(II-V) shell star γ Cassiopeiae, and 3U 1145
061, which seems to correspond to Hen 715 (=SAO 251595), a B1Vne object (cf. Bradt et al., 1977b)

Bradt et al. (1977b) believe that the identification of some X-
ray sources with Be stars may be an indication of the actual existence of the class of X-ray binaries suggested by Maraschi et al.
(1976) where the accreted material is provided by a rapidly rotating
Be star.

Binary nature is also advocated by Joss and Rappaport (1977) to
explain the X-ray bursts sources. They envisage an unevolved OB star
with relatively weak stellar wind that forms a system with a compact
object and the bursts are associated with irregular accretion.

7.   X-RADIATION FROM ALGOL

A very interesting and important new development in X-ray astronomy occurred in 1975 when observations with the SAS-3 satellite in
the 2-6 KeV band detected emission from β Persei (Schnopper et al.,
1976). The measured intensity was 1.6×10^{31} ergs sec-1.

Since Algol is a close binary with no compact component, the SAS-3
discovery unveils another class of X-ray binaries where the generation
mechanism of the X-ray emission may be somehow related to the streams
of matter falling upon the brighter component of the system, but which
do not need the presence of a compact star. The information that could
be gathered in the relevant range of energies from a system like Algol,
will certainly have a bearing on the understanding of the physics of
gas streaming from components of close binaries.

Algol was also detected earlier and later the same year, 1975, with
rockets, in the softer X-ray range of energies from 0.15 to 2 keV

Harnden et al. (1977) find that the fluxes observed are consistent with "a close binary mass exchange model with $\dot{\mathfrak{M}} \approx 10^{-10}\ \mathfrak{M}_\odot\ yr^{-1}$.

8.   THE MASSES OF THE X-RAY BINARIES

One parameter that is very important to help in ascertaining the nature of the compact component in X-ray binaries is the mass. The optical member of the pair is a single-lined object and should normally provide the value of the mass-function of the system. Only when, in addition, the X-ray companion is a pulsating object which undergoes eclipses, and the period of pulsation varies in a manner that reflects the orbital motion of the object, it is possible to derive the individual masses of the two stars. Such is the case of Hercules X-1 where the periodic variation of the pulsation period shows a maximum lengthening of $13\overset{s}{.}19$ in correspondence with the center of the X-eclipse, and a maximum shortening of $13\overset{s}{.}19$ in correspondence with the center of the high state of the radiation level. If the orbit is circular or nearly so, this suggests that the semiamplitude of the velocity curve of the X-ray component is about 169 km sec^{-1} and that the mass-function is nearly $0.86\mathfrak{M}_\odot$. On the other hand, the velocity curves from different spectral lines of the optical companion (cf. Crampton, 1974), although somewhat phase-shifted with respect to the mid- of the X-eclipse, appear to suggest an amplitude of the order of 55 km sec^{-1}. The mass of the optical component would then be about $1.5\mathfrak{M}_\odot$ and that of the X-ray star, about $0.5\mathfrak{M}_\odot$. According to Apparao and Chitre (1976) the "observations are however consistent with $\mathfrak{M}_{opt.} \sim 2\,\mathfrak{M}_\odot$ and $\mathfrak{M}_x \sim 1\,\mathfrak{M}_\odot$ ".

A similar case, as far as the X-ray component is concerned, is provided by Centaurus X-3 where the periodic variation of the pulsation period suggests that the amplitude of the velocity curve of the X-ray component is about 416 km sec^{-1} and that the mass-function of $f(\mathfrak{M}) = 15.6\ \mathfrak{M}_\odot$. However, in this case the optical object is very faint and there is no information available on the orbital motion of Krzemiński's star and, therefore, no direct way of determining the individual masses involved. In a case such as this one, the values of the masses are estimated by analyzing the behavior of the X-intensity with orbital phase, on the assumption that the X-ray source rotates in synchronization with the orbital motion, is in hydrostatic equilibrium, and fills the equipotential lobe. Thus the mass-ratio is derived.

When the compact component is not a pulsating object or when no pulsations have yet been detected, there is no information related to the masses that could be derived from the X-ray observations, and we are only in a position, as we have said, to determine the mass-function from the orbital motion of the optical member of the pair. To derive the mass of the compact component one usually assigns to the optical component the normal mass of a star of the same spectral type and luminosity class and introduces this value into the expression of the mass-function. In this way the masses of the components of Cygnus X-1 have been estimated. The "optical" component of Cygnus X-1 is a well-known single-lined spectroscopic binary, the velocity curve of which yields a semiamplitude of nearly 74 km sec^{-1} and a mass-function of $f(\mathfrak{M}) = 0.23 \, \mathfrak{M}_\odot$ (Brucato and Zappala, 1974). Arguing that the mass of HDE 226868, a 09.7 Iab star, must be larger than $12 \, \mathfrak{M}_\odot$, then the mass of the X-ray source comes out to be larger than about $4 \, \mathfrak{M}_\odot$ and, hence the conclusion that the object must be a black hole.

Trimble et al. (1973) have suggested that HDE 226868 could be a luminous, undermassive object like HZ 22 and then the X-ray companion would have a very small mass. The arguments offered to show that this could not be so are based on the fact that the distance deter-minations suggest that the object is exactly at the distance that corresponds to its luminosity class and not to the luminosity required by the model for a HZ 22-like star. Cheng et al. (1974), however, have pointed out that reddening in the Cygnus region cannot be used for determining distances on account of it being quite irregular.

In Table 8.3 we list the parameters related to the masses of the X-ray binaries that have been derived from the observations.

We have indicated the procedures that are used to estimate the masses when relevant information from only one of the components is available. It seems appropriate now to point out the objections that we should keep in mind when resorting to such procedures (Sahade, 1975; 1976a).

In the first place, the assignment to a star of the mass that would be normal for the spectral type concerned is reasonable if we are dealing with objects that are on the main sequence and have not under-gone any mass loss during their lives. But, even on the main sequence, the well determined masses are not larger than about thirty solar masses (Batten, 1968) and even in some cases that fall within this range, and indeed for larger masses, the individual values are de-rived after some interpretaion of the velocity distribution with phase and are valid as to their order of magnitude and one could only add whether the mass-ratio is larger or smaller than one. Secondly, if we wish to assign a normal mass to an early type giant or supergiant, then the situation is worse because "there are no mass determinations of early type supergiants and the best masses of supergiant stars correspond to spectral type K. The most luminous early-type objects for which there are masses available are a B1 III star, which belongs to a system that does not eclipse, and an A2 II objects; the masses are of the order of five solar masses."

In the third place, would "normal" mass values be valid for members of interacting binaries that have undergone mass loss or have ac-creted mass? The question is certainly even more in order when we are dealing with a binary undergoing a second episode of mass loss. Moreover, over- and under-luminosity are common features among compo-nents of evolved binaries, depending on their present evolutionary stage.

In regard to the question of determining the mass-ratio from the condition of the equipotential lobe being filled by the companion to the compact object, in Chapter 2 we have referred to the results by Sahade and Ringuelet (1970) and we have mentioned the departures of the equipotential surfaces from the classical picture if we intro-duce the effect of radiation pressure, etc. Therefore, this is a method with strong limitations.

Table 8.3.

OBSERVATIONAL PARAMETERS RELATED TO MASSES IN X-RAY BINARIES

X-Ray Component			Optical Component			Masses		Conti's (1977) adopted Masses		Notes
Object	K_x km sec^{-1}	f_x $(\mathfrak{M})$	Object	K_{opt} km sec^{-1}	f_{opt} $(\mathfrak{M})$	$\mathfrak{M}_x$	$\mathfrak{M}_{opt}$	$\mathfrak{M}_x$	$\mathfrak{M}_{opt}$	
3U 0115 −37 SMC X-1	299	10.8	SK 160	19		0.8	12	0.95	15.1	1
3U 0900 −40 Vela XR-1	273	10.5	HD 77581	19		1.4	21	1.6	24	2
3U 1118 −60 Centaurus X-3	416	15.6	V799 Centauri (Krzeminski's Star)	—	—			(1.5)	18	3
3U 1700 −37	—	—	HD 153919	19	0.0023			(1.3)	(27)	4
3U 1956 +35 Cygnus X-1	—	—	HDE 226868	74	0.23			(13)	(21)	5
3U 1653 −35 Hercules X-1	169	0.86	HZ Herculis	~55		0.5	1.5			3

Notes to Table 8.3:

1. Primini, F., Rappaport, S. and Joss, P. C., preprint (1977). The masses quoted in this paper are $\mathfrak{M}_x = 0.8 - 1.8 \, \mathfrak{M}_\odot$, $\mathfrak{M}_{opt} = 15.24 \, \mathfrak{M}_\odot$.
2. Rappaport, S., Joss, P. C. and McClintock, J. E., Astrophys. J. Letters 206, L103 (1976) e = 0.13.
3. cf. Apparao, K. M. W. and Chitre, S. M., Space Sci. Rev. 19 281 (1976).
4. Hammerschlag-Hernsberge, G., preprint (1977). e = 0.16 may be suprious.
5. Brucato, R. J. and Zappala, R. R., Astrophys. J. Letters 189, L71 (1974).

A few years ago Kondo (1974) also critically discussed the assumptions usually adopted to determine the masses of the collapsed components when not enough observational data are available.

9. FURTHER REMARKS

It would fall outside of the scope of this book to attempt a description of the detailed individual behavior of each on of the X-ray binaries. For this we refer the reader to recent review papers like those by Blumenthal and Tucker (1974), Gursky and Schreier (1975), Apparao and Chitre (1976) and Gorenstein and Tucker (1976) and to volumes like Gursky and Ruffini's (1975) and the NASA SP-389 publication which is included several times in the list of references at the end of the book. We would also like to add the paper on "X-ray emission processes in close binary systems" that Pringle presented at the 1976 Eighth Texas Symposium in Boston.

We shall, however, mention briefly the spectral behavior of Cygnus X-1 and Hercules X-1 and will make reference to SS Cygni and to Cygnus X-3.

9.1.   The observations of Cygnus X-1 (Tananbaum et al., 1972) and also of Hercules X-1 (Shulman et al., 1975) have suggested that in both sources there is a variable soft X-ray component. Therefore, when this is strong, the spectra of the sources, that is, the fluxes in terms of the energy, cannot be described by a unique power-law expression.

$$\frac{dN}{de} = A\ E^{-\alpha}\ \text{photons cm}^{-2}\ \text{sec}^{-1}\ \text{keV}^{-1};$$

actually two of them are needed to cover the observed range. The power-law components are different at different times and we can illustrate the case of Cygnus X-1 by quoting from Matterson et al., (1976) the following tabulation:

	1 - 10 keV		10 - 20 keV	
	α	Intensity keV cm^{-2}sec^{-1}	α	Intensity keV cm^{-2}sec^{-1}
September, 1970	2.6	10	2.4	3
December 21, 1970	3.8	50	1.6	20
April, 1971 } Dec.,1971-Jan.,1972 }	1.5	5	1.5	30

In their review paper on soft X-ray sources, Gorenstein and Tucker (1976) mention the models that have been proposed for explaining the behavior of Cygnus X-1 and Hercules X-1. It is considered that the

soft X-ray radiation arises from the accretion disk and that the variations in the emitted flux are perhaps related to variations in the
rate with which matter is accreted onto the X-ray component.

9.2.   Soft X-ray from SS Cygni have been also reported (Rappaport
et al., 1974; Heise et al., 1975), in agreement with what one would
expect if the accreting component of the system is a white dwarf.
The luminosity of SS Cygni in the soft X-ray range corresponds to
10^{33} ergs sec^{-1} and Pringle's (1977) computations suggest that the
X-ray radiation in dwarf novae comes from the "boundary layer where
the accretion disk grazes the surface of the white dwarf".

9.3.   Cygnus X-3 is one of the X-ray binaries which have been observed in various ranges of energy simultaneously. Observations were
carried in X-ray, infrared and radio in 1973 (Mason et al., 1976) and
one important piece of information derived was that at times the
infrared measurements are characterized by the 4.8-hour periodic
structure which is typical of Cygnus X-3. The same periodicity is
displayed by the observations in γ-ray radiation (energies greater
than 100 MeV) detected with the SAS-2 satellite in 1973 (Lamb et al.,
1976). It is interesting to quote from a paper by Apparao (1977)
the energy radiated by Cygnus X-3 in the different ranges. They are
given in the accompanying tabulation:

Radio	$\sim 10^{31} - 10^{34}$	ergs sec^{-1}
Infrared	$\sim 10^{36}$	ergs sec^{-1}
X-ray	$\sim 1-4^{37}$	ergs sec^{-1}
γ-ray	$\sim 10^{37}$	ergs sec^{-1}

Space astronomy is at its beginning and has yielded already very
important and exciting developments. The increase in sensitivity and
the simultaneous coverage in various ranges of energy will be bringing
about further advances and certainly a better understanding of the
actual evolutionary history of the interacting binaries. Moreover,
the X-ray binaries are perhaps the only object that could disclose
the existence of a black hole component in a close pair and the discovery of one of them would certainly be another major breakthrough
resulting from the utilization of space technology in Astronomy.

CHAPTER 9

GROUPS OF SPECIAL INTEREST

1.   RS CANUM VENATICORUM SYSTEMS

1.1.   The concept of this group as a special type of close binary
with specific characteristics has developed rather gradually. Although certain individual systems have long presented problems in
interpretation, it is only in recent years that a distinct class has
been recognized. Not only do the individual systems present puzzles
but the existence of the class as a whole is difficult to explain.
Their space density is very high for binary stars; in fact, it seems
to be at least comparable to the W Ursae Majoris systems which for
many years have been thought to be most common type of close binary
systems. The evolutionary problem of why there should be so many
systems of this type has no clear answer at the time of this writing
All mechanisms suggested to date have raised questions which so far
have not been completely solved. Recently Hall (1976) has given a
comprehensive survey of the observed properties. Hence, we shall·on
summarize these and then discuss in more detail the two systems whicl
have been most thoroughly studied.

The main properties were first proposed formally by Oliver (1974)
but, as he recognized, many of the suggested properties were not pre
sent in all systems. Since this proposal, other characteristics hav
been noted. The cooler component of these systems is a subgiant of
class K0 or very close thereto; the hotter component is a F or G sta:
either on the main sequence or slightly above it. The two are well
separated compared to their radii and in all cases form a "detached"
system - a feature which makes still more curious some of their othe
properties. Despite the difference in spectral types, the individua
masses usually are nearly equal. The combined masses come to betweer
$1.0\,\mathfrak{M}_\odot$ and $4.0\,\mathfrak{M}_\odot$ with the majority falling between $1.75\,\mathfrak{M}_\odot$ and $3.0\,\mathfrak{M}_\odot$.
There is indication of strong chromospheric activity as shown by H
and K lines in emission. Further, radio emission has been detected
from a few including one which is only ninth magnitude optically.
The orbital period of those known at present lies between 1.98 days
(AR Lacertae) and 10.72 days (RV Librae). Sudden, unpredictable
period changes, including some relatively large ones, are normal for
these systems, although usually these are extremely rare in detached
binaries. In addition to the eclipse effects, the most carefully
studied light curves show waves which seem to migrate progressively
along the light curve toward decreasing orbital phase. In addition

to these changes, irregular light variations are present. Both UV
and IR excess have been observed. These systems have not been found
in regions of star formation; none have yet been discovered in clusters.

We emphasize that not all these characteristics have been found in
every system. As a working definition, Hall proposes the following:
(1) orbital period from 1 day to 2 weeks, (2) hotter component of
spectral class F or G and luminosity class IV or V, and (3) strong H
and K emission outside eclipse. We will now discuss in some detail
the two most extensively studied examples of the class, AR Lacertae
and RS Canum Venaticorum.

1.2. The light variability of AR Lacertae, but not its nature or
period was discovered by Miss Leavitt (1903) and confirmed by Wendell
(1907). Various other observers, on the basis of scattered observations, either found no variability at all or found a range much less
than the 0.6 mag. originally reported. Finally, a series of observations by Jacchia (1929a, 1929b) and Loreta (1929, 1930) established
the fact that this was an eclipsing system with flat-bottomed minima
and a shallow secondary. The duration of eclipse and of totality
were each determined, and a period of 1.98 days was found. The period
of almost exactly two days; this and the small out-of-eclipse variation
explains, of course, why earlier observers had trouble confirming the
variability. The period and general nature of the light curve was
confirmed by several later sets of photographic observations and one
photometric solution was made (Schneller and Plaut, 1932). Harper
(1933) presented the first spectrographic orbit. He found a small
eccentricity of orbit (e = 0.041); his solution, using the inclination
of 86° as determined from the light curve, gave masses 1.41$\mathfrak{M}_\odot$ and
1.42$\mathfrak{M}_\odot$, respectively. Harper noted the disagreement between equal
masses and nearly identical spectral types on the one hand, and on the
other a ratio of surface brightnesses (from the light curve) of nearly
three to one. He pointed out, however, that secondary minimum was not
well observed. Indeed, Schneller and Plaut had noted difficulties with
the duration of secondary.

Wyse (1934) classified the spectra as K0 and G5, with the K0 component being fainter on the spectrographic emulsion but probably brighter
visually. He noted that the K0 component had sharp H and K emission
lines, superposed on broad absorption, and appeared intermediate beween
a giant and a dwarf both in density and in luminosity. Various other
observers presented light curves without recognizing discrepancies
larger than those usually present when comparing different sets of
photometric estimates. However, in rediscussing all the observations
through 1975, Hall et al. (1976) finds both a persistent wave-like
feature and an additional intrinsic light variation.

Himpel (1936), using filters and a wedge photometer, reported a hump
during primary. Wood (1946) published the first photoelectric light
curve. He called attention to a sudden period change and the probabil-
ity of intrinsic variation of one component. The comparison star he
used - always a matter of suspicion - was checked regularly and apparently was quite constant during the observing intervals, although this
star has later been reported to be variable by Blanco and Catalano
(1970). In the 1950 study of period changes in general this was found
to be the only case of a detached system showing irregular changes.
Kron (1947) noted irregularities in the partial phases and first suggested an explanation based on "star spots" on the photosphere of the
eclipsed component. Many spectroscopic and photometric papers were

published during the next twenty years, including a note by Struve
(1952) discussing the spectra in the light of a "turbulent-spot" hypothesis.

Since about 1968, the system has attracted considerable attention,
and has been extensively observed. At the time of this writing, the
most recently published thorough study is that of Chambliss (1976)
who discussed a series of about 2000 three-color photoelectric observations. He finds intrinsic variation of about 0.04 magnitude. Smaller scatter in primary minimum than in secondary suggests this is
chiefly associated with the hotter component. Radii of 1.54 $R_\odot$ and
2.81 $R_\odot$ are found although the masses are identical at 1.35 $\mathfrak{M}_\odot$. The
relative radii are in satisfactory agreement with earlier values and
indicated well separated components with $a_1 = 0.17$ and $a_2 = 0.32$.
However many problems remain unsolved and extensive future observations are urged. Hall et al. (1976) studied 18 light curves from
1926 to 1974 and times of minima from 1900 to 1972. They report a migrating wave of varying amplitude and suggest period changes occur
when the minimum of this is between $0^P.25$ and $0^P.76$. The evidence is
not conclusive, and it will be interesting to see if their prediction
of a change in the vicinity of 1974 is in fact observed. At present
the statement made in 1946 - "The present knowledge of this star is
far from satisfactory" is still valid.

An exciting recent development has been the detection of radio sig-
nals. Hjellming and Blankenship (1973) first reported variable radio
emission from observations made in February 1973. On October 16-18,
1973, Gibson and Hjellming (1974) observed a relatively strong radio
flare of the "Algol-type" at 2695 and 8085 MHz. This shows a flat
radio spectrum before the flare with higher flux levels at higher frequencies during the flare followed by a decay of the flare event until
again a relatively flat spectrum is reached at low level of flux.
Compared to Algol (the only other radio binary sufficiently observed
at the time to make comparison meanginful), other similarities were
found. The duration of the flare was approximately 24 hours, compared
to an average value of 16 hours for typical Algol flares. The rise
time for AR Lacertae was "somewhat more rapid" and the spectral index
increased "much more rapidly". In both cases, the onset of increased
radio emission came on time scales of minutes. This suggests a sudden
reenergization of radio emitting regions of sizes of the order of tens
of light minutes or less. Mullan (1974) has suggested that the radio
bursts are associated with regions surrounding star spots. This system
is under reasonably constant observation both in the optical and radio
ranges, but nearly 75 years after its variability was announced, obviously there is still much to be explained.

1.3.   The other frequently observed system in this class is the one
for which it is named, RS Canum Venaticorum. This too has a long
history and it was discovered and correctly identified as Algol-type
by Ceraski (1914). Note that while we have mentioned distinct characteristics which enable these systems to be differentiated as a special class, these require further study than merely finding the shape
of the light curve; the conventional and convenient customs of making
an initial classification of either Algol, W Ursae Majoris, and β Lyrae
types can usually be done even from light curves based on estimates.
and will probably continue to be so used. However, systems showing
Algol-like light curves are by no means always among those characterized as "Algol-type" systems.

The work of several early observers, among whom C. Hoffmeister
and Nijland were the most assiduous, established the nature of the
light curve of RS Canum Venaticorum and determined the light elements.
The Mount Wilson annual report for 1923 mentioned that observations
at primary showed the spectral type to be K0. Sitterly (1928) made
the first photometric solution using both visual and photographic
estimates. He noted asymmetry both in primary and between the eclipses,
and found that the fainter star was redder than the brighter. With
unpublished spectrographic elements supplied by Joy, he computed masses
of 1.88 $\mathfrak{M}_\odot$ and 1.74 $\mathfrak{M}_\odot$ and a density for the fainter component of only
0.011 $\mathfrak{M}_\odot$. He concluded that the brighter component was a "fairly normal" F dwarf while the fainter might be called a sub-giant. A preliminary report by Joy had reported spectral types of F3 and K0 with
the F3 being stronger. The masses were nearly equal and about the
solar mass. The conclusions, however, all rested on three plates. Later,
on the basis of 35 spectrograms, Joy (1930) classified the components
as F4n and dG8. Both were dwarfs but the fainter was exceptionally
large and massive. Schneller (1928) also presented an extensive discussion based on Joy's spectrographic results and a photographic light
curve obtained with more than usual precautions. He found the F3
component to be a dwarf and the K0 between a giant and a dwarf. Both
were too bright for their masses.

For reasons of space, we cannot discuss in detail the large amount
of work done during the next forty years.

The variability of the period was noted and studied. The presence
of CaII in emission was observed in the spectra. Intrinsic variations
in the light curve were established beyond question, including an unexplained increase in the brightness during totality without a corresponding increase outside eclipse. While the unusual light changes made
a precise solution impossible, a reasonably clear picture had emerged
of a well separated system with the combined radii equalling only about
one third of the distance between centers and each component well removed from the critical zero-velocity surface. The masses were approximately equal and about 1.4 $\mathfrak{M}_\odot$. Approximate radii in solar units
were 2 for the hotter and 4.5 for the cooler star; these were subject
to considerable uncertainty. We now consider some of the results found
in the modern era.

In 1967, Catalano and Rodono (1967) reported on a series of photoelectric observations made in 1965-66. They confirmed the luminosity
variations reported for secondary star; other distortions of the light
curve were discussed and attributed to a ring around the primary component whose equatorial plane was inclined with respect to the orbital
plane. This, they suggested caused the displacement of the secondary
eclipse which had been reported by various observers and attributed
to orbital eccentricity. Thus, the observed period changes could not
be attributed to apsidal motion. Nelson and Duckworth (1968) also
reported variations in the light curve; these included significant
changes in the relative heights of maxima. Catalano and Rodono, (1969)
continuing their observations, added further details. They found the
fluctuations were fairly regular; outside of eclipse, they could be
represented by $L = L_0 - \Delta L \cos \{\phi - \theta(t)\}$ where ϕ was the phase angle
and θ gives the phase angle of the minimum of the wave-like distortion
relative to primary minimum. A nearly sinusiodal variation outside
of eclipses showed a period of 7-10 years. In addition to explaining
the present nature of the system, much discussion has been given to
explaining the evolutionary questions posed by this type of system.

Hall (1972) prepared a model to explain the major problems connected with the system. He lists these as (1) the wave-like distortion which distorts the light curve outside the eclipse; (2) the fact
that this maintains its shape while migrating year by year towards
decreasing orbital phases; (3) the migration rate is not uniform; (4)
the depth of secondary is about half that to be expected from the relative surface brightness as deduced from the color indices; (5) the
variation with time of the depth of primary; (6) the variation of the
location of secondary minimum relative to the adjoining primaries;
(7) the amplitude of the travelling wave has varied from V = 0.2 mag.
to amounts difficult to detect with certainty, and (8) the large and
seemingly erratic fluctions of period even though each component is
well inside its limiting lobe. Earlier attempted explanations of
some of these had included tidal lag, a third body, absorbing circumstellar material, the pulsation of one component, a rotation of the
line of apsides, and the tilted ring model previously mentioned.
Catalano and Rodono (1969) had very clearly demonstrated that the
cooler component must be responsible for the wave by their observations of variations in the depths of the primary (total) eclipse.

Hall's model called for a region of "tremendous sunspot activity"
which darkens one side of the cooler component; the spots (as in the
case of the sun) are confined to a region within 30° of the equator.
Differential rotation (also like the sun) then produces the migration
of the wave-like distortion. A $23\frac{1}{2}$ year starspot cycle in the cooler
component accounts for the variable amplitude of the wave; during the
cycle, the spots form at different latitudes (as with solar spots),
and this can account for the non-uniformity in the migration rate.
The increases and decreases of period were explained by continuous
ejection from a migrating active region. The assumed velocities of
ejection were 3000 km/sec which implied a mass loss of $10^{-6}m_{\odot}$ /yr.
Later Arnold and Hall (1973) noted that the active regions should be
identified with the brighter regions of the cooler component, not the
fainter hemisphere as originally suggested. Many problems still remain with this model, as Hall has emphasized. For example, how do
we explain physically why sunspot activity persistently prefers one
hemisphere despite the disruptive influences of differential rotation?

Catalano and Rodono (1974) have argued against the existence of
the short period fluctuations discussed by Hall on the grounds that
the asymmetry shown by the minima and its variations with the position of the migrating wave mean that precise times of minimum are
difficult to obtain. Hall (1975b) noted that three different systems
of this type (RS Canum Venaticorum, SS Camelopardis, and CG Cygni)
show the same correlation between the (O-C) variations of minima and
the outside eclipse wave migration even when the effects of light
curve asymmetry have been removed. Oliver (1975) has discussed the
tilted ring model from the point of view of celestial mechanics, and
has shown that it is extremely difficult to explain its present existence in a quantitative way and even more so to explain its continu-
ation through the years the system has been under observation. However, whatever the final solution, the long series of excellent observations by the Catania observers will play an important role.

Observational work so recent that its significance has not been
fully considered as yet, include the polarization observations of
Pfieffer and Koch (1973, 1977), the correlation reported by Weiler
(1975) between the H, K, and $H\alpha$ emission and the minimum of the wave
like distortion, and the infrared excess reportedly found at the

Edinburgh Observatory (Quart. J. R. ast. Soc. 17, 147, 1976). The
only results from the radio observations to date are the upper limit,
at least at one epoch, reported by Altenhoff et al. (1976).

1.4.   The evolution of these systems remains a problem. Hall (1972)
pointed out the present state must be explained either by post-main
sequence evolution (1) before mass exchange or (2) after mass exchange, or (3) pre-main sequence evolution. He felt that the latter
could account for the observed properties, while the others encoun-
tered difficulties. He notes that the cooler components show some
of the properties of the T Tauri stars especially in the H and K
emission. Atkins and Hall (1972) made infrared (JHKL) photometry of
seven of these systems. Of the six adequately observed, five showed
infrared excess. The average excess of 0.5 magnitude was thought to
be almost certainly significant and associated with the cooler component. The wave-like distortion was also observed at these wavelengths.
However, later Hall (1975c) discussed WW Draconis, a system of this
class, and because of the presence of its less massive F8V visual
companion, concluded it could not be in pre-main sequence contraction.
Bopp and Fekel (1976) drew similar conclusions from the visual companion to HR 1099. It is extremely difficult to imagine different
stages of evolution for so homogeneous a class of stars.

    At the time of the 1975 Cambridge symposium (Symposium No. 73 of
the International Astronomical Union), Biermann and Hall suggested
(by process of elimination) that these systems are formed by fission
of rapidly rotating main sequence stars. However, this left serious
problems connected with the amount of angular momentum and the rotation of the core. At the same symposium, Ulrich suggested high mass
transfer. He, however, found the light curve irregularities difficult to explain. More recently, Popper and Ulrich (1977) have proposed that these evolve from "nonemission" binaries with mild mass
exchange and probably mass loss. Their arguments are based on the
distribution in the mass-color, mass-radius, and color-luminosity
planes. However, at the present time, it is still somewhat difficult
to explain the existence of these systems.

2.   WOLF-RAYET BINARIES

    The existence of such peculiar objects as the Wolf-Rayet stars has
been known since 1867. Two French astronomers, Charles Joseph Etienne
Wolf and Georges Antoine Pons Rayet (1867) observing at the Paris Observatory with a visual spectroscope, discovered three stars in the
constellation of Cygnus, namely, HD 191765, HD 192103 and HD 192641,
that displayed very peculiar spectra characterized by highly broadened
emission lines superimposed on a weak continuum. Four years after-
wards Lorenzo Respighi discovered that γ_2 Velorum is also a Wolf-Rayet
star, the brightest in the sky.

    It was almost three quarters of a century later that for the first
time a Wolf-Rayet star, also in the constellation of Cygnus, was found
by Wilson (1939) to be a spectroscopic binary with a period a little
over four days. This important discovery was followed one year later
by Gaposchkin's (1940) announcement that this star, now known as V444
Cygni (=HD 193576) was an eclipsing object and derived light elements
(Gaposchkin, 1941). The discovery of the first Wolf-Rayet binary
created the hope that finally the door to the knowledge of the nature
and physical properties of this group of objects was open.

Since the discovery of V444 Cygni as a close binary system several additional Wolf-Rayet binaries have been found, and the presently known cases are listed in Table 9.2.1.

Table 9.2.1.

WOLF–RAYET BINARIES

Object		Period (days)	Spectral type	Remarks
HD 68273	γ_2 Velorum	78.5	WC8 + O8I	
90657		6.5	WN5 + O6	
92740			WN7	
113904	θ Muscae	18.3	WC6 + O9.5	eclipsing?
152270		8.8	WC7 + O5–8	
168206	CV Serpentis	29.7	WC8 + BO:	
186943		9.6	WN4 + B	
190918		85.0	WN4 + O9I	
193576	V444 Cygni	4.2	WN5 + O6	eclipsing $A_1^{ptg} = 0.30$ $A_2^{ptg} = 0.14$
193928		21.6	WN6	
197406		4.3	WN7	
HD 211853		6.7	WN6 + O6I	eclipsing?
MR 114	CX Cephei	2.1	WN5	eclipsing $A_1^{ptg} = 0.12$ $A_2^{ptg} = 0.04$
HD 214419	CQ Cephei	1.6	WN6	eclipsing $A_1^v = 0.47$ $A_2^v = 0.39$

The Wolf-Rayet binaries may belong to either one of the two parallel sequences in which the group is divided, namely, the carbon (WC)
and nitrogen (WN) sequences; in the latter we have two subsequences,
depending on whether the emissions are broad or relatively narrow.
Their spectra can be described as the juxtaposition of three sets of
spectral lines: Firstly, we have the characteristic emission spectrum with very broad lines of C and O (in the carbon sequence) or N
(in the nitrogen sequence) and H, He, Si and other elements in several
ionization stages; some of these emissions may have P Cygni-type profiles;

Secondly, we have a stellar absorption spectrum, usually corresponding to O or B spectral type; and

Thirdly, we have an absorption spectrum that suggests diluted radiation and in the photographic and near infrared regions shows best
in the triplet lines of He I $\lambda\lambda$ 3888 and 10830 Å.

Furthermore, some of the broad emission lines, which were usually
called "bands" just because of their width, display a superimposed
narrow emission. This narrow emission has been found in objects like
V444 Cygni and γ_2 Velorum and shows principally in HeII λ 4686.

The broad emission spectrum displays the correlations that are
characteristics of the Wolf-Rayet objects (cf. Kuhi, 1968), namely,
a) the higher the ionization, the narrower the lines; b) in the WC
sequence, the earlier the spectral type, the broader the lines, and
c) the WN subsequence with relatively narrow lines appears to be
largely formed by close binary systems. The emission profiles, are
either Gaussian or flat-topped, as we mentioned in Chapter 5 (cf.
Beals, 1968). We should add that the P Cygni profiles that sometimes
are present lend support to the interpretation of Wolf-Rayet spectra
first advanced by Beals in terms of matter being ejected by the stellar object.

Normally the stellar absorption spectrum yields a radial velocity
distribution that shows a large scatter and suggests a curve 180°
out of phase relative to the velocity curve from the emission lines.
This fact led to the belief that whenever we had a Wolf-Rayet star
displaying absorption lines of H and other elements (not the violet
edges of the P Cygni profiles nor the lines that indicate dilution
effects) we were dealing with a binary system where these absorptions
arose in the atmospheres of the companions to the Wolf-Rayet objects.
This belief was specially useful for statistical studies of Wolf-
Rayet binaries because it was sufficient to examine low dispersion
spectra when the star was too faint for taking slit spectrograms and
measuring radial velocities. However, the Wolf-Rayet binary HD 92740
(Niemelä, 1973) did conform to the above description and yet the
velocity curves from the H absorptions were in phase with those from
the emission features. This suggests that the H absorptions arise
from the Wolf-Rayet component of the system and, therefore, raises
the possibility that the H velocities in other Wolf-Rayet binaries
may not reflect purely the orbital motion of the companion to the
Wolf-Rayet star but yield values that result from the blending of the
lines of the two components of the systems.

As for the absorption lines that suggest diluted radiation, we
have already mentioned in Chapter 5 that they yield velocities of
approach of over 1000 km sec^{-1}; in a number of objects they are associated with flat-topped emissions. We have also mentioned that, at

least in some cases, HeI λ 3888 displays multiple components. The strongest component is the more violet-displaced one and there seems to be a phase-dependence on the appearance of the feature (cf. the discussion in Bappu and Sahade, 1973). In Chapter 5 and 8 we have made reference to the intercombination line CIII] λ 1909 Å found in the ultraviolet spectrum of γ_2 Velorum (WC8) and that this line suggests a velocity of approach as large as the one suggested by the triplet lines of HeI. These lines must be formed in the outermost layers of the "outer envelope" of the systems and hence, the relevant electron densities must be of the order of 10^{10} cm^{-3}, as mentioned in Chapter 5.

It is tempting to add, for the sake of completeness, that in a WN object, although it has not been proven yet to be a binary, namely HD 50896 (WN5), the ultraviolet spectrum displays the line of NIV λ 1488 Å (Smith, 1972).

Because of the remarkable spectra that are associated with the Wolf-Rayet stars, and the fact that many peculiar objects have been shown to be interacting binaries (which is also the case for a number of the Wolf-Rayets) the question has arisen many times as to whether binary nature is a necessary condition for their existence. The fact that some Wolf-Rayet stars do not give evidence of orbital motion or of the presence of a companion are not arguments against that possibility because the distribution of the orbital planes in space should be at random and, therefore, we should find cases where no orbital motion can be detected. Regarding the companion, it may not be apparent in the photographic regions, for instance, but it may show in the ultraviolet, as it is the case in HD 92740. In this binary the ultraviolet spectrum taken from the Skylab which extends to about 1300 Å, shows "a strong continuum above which the emissions are barely visible" (Henize et al., 1975) suggesting the presence of the companion of which no lines have been detected so far in the blue region of the spectrum.

De Monteagudo and Sahade (1970, 1971; see also Bappu and Sahade, 1973) have suggested that the periodic changes in the intensity and structure of the violet-displaced diluted absorption lines of HeI λ 3888 and/or the periodic changes in the V/R intensities of the components of the double emissions of H and HeI should be good indicators of the binary nature of a Wolf-Rayet object. Plates of γ_2 Velorum taken in 1919 by Perrine and in the interval 1948-1961 by Sahade suggests that we are dealing with a phase-dependent phenomenon, a fact that appears to be also true in V444 Cygni. The red component of the emission becomes stronger than the violet at the quadrature that follows the conjunction at which the Wolf-Rayet star is in front.

Nine of the fourteen systems listed in Table 9.2.1 are double-lined in the photographic region.

The Wolf-Rayet stars are difficult to measure for radial velocity on account of the nature of the spectral lines. In general, the emissions suggest systemic velocities which are red-shifted by different amounts relative to the systemic velocities that are suggested by the radial velocities from the stellar absorption lines. In addition, the amplitudes of the velocity curve from different emission features may be different, and even the velocity distribution may suggest difference in trends.

In regard to the redshift of the emission lines, Niemelä (1973) seems to have found evidence that, at least in the case of He II λ 4686 in HD 92740, the redshift is due to the effect of an undetected violet-

displaced absorption component. If this is so and if the atmosphere
of a Wolf-Rayet star is stratified and there is a velocity field, as
correlation a) shown by the emission lines appears to indicate, then
an interpretation of the different redshifts may require knowledge
of the structure of the atmosphere.

In HD 92740 (Niemelä, 1973) the Balmer progression suggests that
the Wolf-Rayet atmosphere accelerates outwards, a fact which would
normally imply that the temperature decreases as we go away from the
star. Kuhi (1973) has given strong arguments that suggest that, at
least in the WC sequence, "the temperature must decrease radially outwards from the star". Therefore, the correlation between the velocity
from the violet-displaced absorptions and the ionization potential of
the corresponding ions, in the sense that the higher the ionization
potential the smaller the velocity, found by Bappu (1973) in several
Wolf-Rayet stars, by Seggewiss (1974) in HD 151932, and by Niemelä
(1976) in the binary HD 90657, are additional arguments that support
the conclusion that the atmosphere of Wolf-Rayet stars is accelerated
outwards.

If such is the structure of the atmosphere of the objects we are
dealing with, then we can infer that

a) if the redshifts are the result of, in many cases undetected
violet absorptions, then there should be a correlation between the
amount of the redshift and the ionization potential of the corresponding ion;

b) the orbital elements of the Wolf-Rayet component should perhaps
be derived on the basis of the velocity curve from the emission which
would originate in the layers closer to the star itself.

In some cases the observations appear to suggest that the matter
which is ejected by the Wolf-Rayet star and gives rise to the accelerated envelope where the emissions originate, mostly comes from the
hemisphere that is away from the companion (cf. Cowley et al., 1971).

The best studied of the Wolf-Rayet binaries is V444 Cygni for which
there are photoelectric light curves in λλ 3550 (Hiltner, 1949) 4500
(Kron and Gordon, 1943) and 7200 (Kron and Gordon, 1950) and a good
spectrographic investigation by Münch (1950).

The light curves (Kron and Gordon, 1950) showed, in the three wavelengths, a primary minimum twice as long as the secondary, suggesting
that the Wolf-Rayet component is surrounded by an electron-scattering
envelope that extends to a distance of about eight or nine times the
radius of the star itself. Furthermore, these observations suggested
that there also exists an "inner luminosity" envelope of a radius a
little over three times the star's radius. This latter envelope is
effective at secondary minimum, when the Wolf-Rayet component is behind the O6 companion. The ratio of the radius of the O star to that
of the Wolf-Rayet component appears to be about 4.5.

It is interesting to mention at this point that the measurements
of γ_2 Velorum made by Brown et al. (1970) with a stellar intensity
interferometer at the Narrabi Observatory, in the continuum at λ 4430
and in the CIII-IV λ 4650 emission, suggested that the size of the
emitting region is about 4.5 times larger than the size of the region
responsible for the continuum.

Unfortunately, the eclipses of V444 Cygni are partial and since

there is no good determination of the luminosity ratio of the two
members of the pair, we cannot infer the actual dimensions of the
stars nor their densities (Sahade, 1965). The masses by Münch are,
on the other hand,

$$\mathfrak{M}_{WR} = 10.3 \; \mathfrak{M}_\odot \quad \text{and} \quad \mathfrak{M}_\odot = 26.1 \; \mathfrak{M}_\odot$$

The reliability of the rest of the masses is, in general, low and
furthermore, the inclinations of the orbits are not known. As Kuhi
(1973) puts it, the masses of the Wolf-Rayet stars "are not at all
well determined".

In the discussions related to Wolf-Rayet stars the mass is assumed
to be of the order of 10 $\mathfrak{M}_\odot$, but there are indications that in some
systems it may be larger. So far, the results have suggested in
every case that the mass of the Wolf-Rayet component is smaller than
the mass of the companion and if this can be ascertained in view
of the observations of HD 92740, it must be quite a significant observational fact. Because of this feature, the luminosities involved,
the mass assigned to a Wolf-Rayet star and the fact that so far it
has been accepted that we are dealing with H-deficient objects, our
group of binaries is thought as resulting from "mass exchange", and
the Wolf-Rayet components, as objects in their helium-burning stage
of evolution.

There are observations that suggest that there is a large amount
of matter between the two components of the Wolf-Rayet binaries. It
was first shown by the light curve of CQ Cephei in λ 4686 of HeII
(Hiltner, 1950), which had its maximum at conjunctions and it minimum
at quadratures. Let us mention two further observations that point
in the same direction. One is the narrow emission that appears superimposed upon some of the broad ones, principally HeII λ 4686, as we
have indicated when describing the spectrum of a Wolf-Rayet binary.
In V444 Cygni such a narrow emission undergo periodic shifts with
phase and suggests that it comes from matter between the two components which is moving from the Wolf-Rayet towards the O companion
(Sahade, 1958b).

The second point to mention is related to the observations of the
Wolf-Rayet binary CV Serpentis. This star was discovered to be an
eclipsing variable by Gaposchkin (1949), the eclipses were confirmed
and found deeper by Hjellming and Hiltner (1963), and, then, less
than a decade later Kuhi and Schweizer (1970) could not detect any.
One possible explanation is that the eclipses observed earlier were
not of the stars but of matter between the stars (Cowley et al.,
1971).

The narrow emission superimposed upon the broad HeII λ 4688 in
V444 Cygni is responsible for the asymmetry, and the phase-dependent
changes in it, reported for that line by Wilson (1940). In γ_2 Velorum,
Sahade (1958b) also found a narrow emission superimposed upon HeII
λ 4686, but its behavior appear at the moment to suggest that its
origin should be explained differently than in V444 Cygni. At any
rate, the narrow emission and its changes might be responsible, at
least partly, for the non-periodic variations that have been detected
in γ_2 Velorum in the intensity of HeII λ 4686. Figure 9.2.1. shows
the results of the photoelectric observations of this line made by

R. R. D. Austin at the Mount John Observatory in New Zealand.

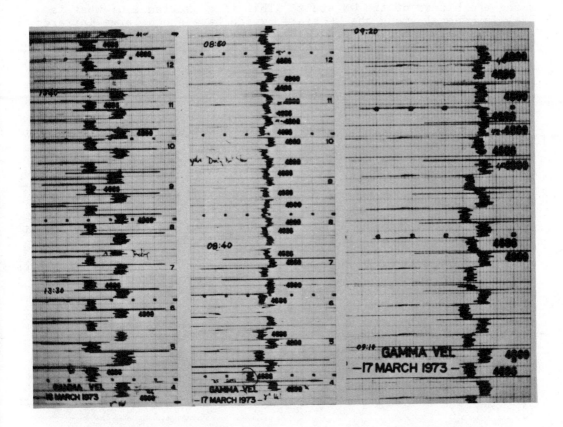

Fig. 9.2.1. Changes in the intensity of HeII λ 4686 in γ_2 Velorum. The
photoelectric observations are made alternatively on the continuum at
4800 Å and on the emission line at 4686 Å. The tracings show two consective
nights in March, 1973; time increases from the bottom of the tracing to the
top and each occupies 20 minutes total runs.
Note that on March 17 the 4686 deflections were noticeably higher than the
night before, during some 5 minutes the emission line intensity increases
to the brightness of the continuum, then it decreases to flare up again
in a time interval of a few minutes.
(Observations made at the Mount John Observatory in New Zealand by
R. R. D. Austin.)

The infrared photometry of the Wolf-Rayet stars at 1.6 and 2.2 μ (Allen et al.,1972; Allen and Porter, 1973) has found no strong infrared excesses among the WN binaries; furthermore, the two measured southern WC binaries, namely, γ₂ Velorum and θ Muscae, are bluer than any of the WC stars observed in the survey, and actually they are among the bluest of the WN stars. This is at variance of what is true in general among the WC stars, which have shown large infrared excesses. Allen and Porter (1973) believed that the reason for this must lie in the fact that the companion stars, both in γ2 Velorum and in θ Muscae, are very luminous objects and "to these, perhaps, should be attributed the destruction of the material normally producing the infrared excess" In the ultraviolet the luminosity ratio of the two components of γ₂ Velorum has been found to be one at 2100 and 2500 Å (van der Hucht, 1975).

The photometry carried farther in the infrared, from 2.3 μ to about 10 μ by Hackwell et al. (1974), that included several northern Wolf-Rayet binaries, seems to suggest, however, that these binaries have infrared excesses which are small or null at shorter wavelengths and, in most cases, appear to increase with wavelength from 4.9 μ on. Such infrared excesses can be understood in terms of free-free emission from a circumstellar plasma.

Since all evolutionary considerations and models of the system with Wolf-Rayet components recur to the drawing of the zero-velocity surfaces, it seems appropriate to show in Fig. 9.2.2. how such surfaces look like when one takes into account the effect of the radiation pressure from the two components. The application of Zorec's (1976) results to the case of V444 Cygni has been done by Niemelä (1977).

There are many questions that need an answer before we fully understand the Wolf-Rayet stars. Even the problem as to whether they are all binaries is still open. Now there are observations of Wolf-Rayet stars that extend from the far ultraviolet through the radio region of the spectrum and it is now possible to gather information about the behavior of the objects in different layers of the extended gaseous envelopes in which they are embedded. This opens up a way for further advances in our knowledge of this fascinating group of objects.

For additional information on the Wolf-Rayet stars we refer the reader to the volume of the proceedings of the IAU Symposium on Wolf-Rayet and High Temperature Stars, held in Buenos Aires in 1971, and to Special Publication No. 307 of the U. S. National Bureau of Standards (1968 Symposium on Wolf-Rayet Stars, in Boulder).

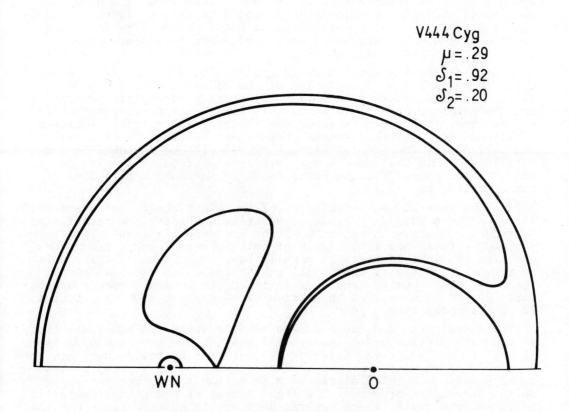

Fig. 9.2.2. The Roche equipotential surfaces in V444 Cygni
when considering the effect of the radiation pressure from
the two components.

3.   DWARF NOVAE

3.1.   It is currently believed that, with the exception of some supernovae, all explosive variables are interacting binary systems.   Indeed, attempts have been made to explain at least one type of supernova explosion as originating in such a system.   In one such model,
it is assumed that the system is composed of a white dwarf and a star
in the process of evolution toward the red giant stage.   The envelope
of the latter will eventually expand to include the white dwarf which,
because of friction, will spiral in closer and closer to the nucleus
of the red star with the final result being a supernova explosion.   Another suggested possibility is that the white dwarf accretes matter
from its companion until it is driven over its mass limit.   These ideas
have developed in some detail by various authors, e.g., Sparks and
Stecker (1974), Gursky (1976), van den Heuvel (1976).

Anything approaching a complete treatment of explosive binaries
would require at least a book of its own. Here we will concentrate
on some of the major developments in the U Geminorum and the Z Camelopardalis stars, usually designated as dwarf novae. The following
section will discuss classical novae and, when pertinent, some features of recurrent novae.

Glasby (1970) has written an informative book on the dwarf novae
which is an excellent description of the older material. Warner
(1976) has produced an extremely comprehensive description of the
observational material related to this class of explosive variables
and Faulkner (1974) has reviewed the theoretical aspects. Robinson
(1976) has included them in his discussion of structure of cataclysmic variables. The general picture, which will be reviewed in more
detail later, is that of a binary system comprised of a white dwarf
and a cooler component; the white dwarf is surrounded by a disk of
circumstellar material.

3.2.   The general characteristics of the light changes of dwarf novae
have long been known. Periods of relative stability are interrupted
by sudden outbursts in which the light increases in a nova-like fashion, but with an amplitude range of only between 2 and 6 magnitudes.
The time of light increase is usually less than 24 hours. There is
no regular periodicity but in all cases studied to date, a "time
scale" or rough period can be established; in general these range between 30 and 50 days, but in extreme cases can run from as little as
12 days to more than a year.

The U Geminorum and the Z Camelopardalis stars are sometimes listed
as separate classes under the general heading of "nova and nova-like.
variables" or "eruptive variables:, but in recent years there has been
a strong tendency to consider them as subclasses of a group called
dwarf novae. The chief distinction is a standstill frequently found
in the descending branch of the light curves of the Z Camelopardalis
stars which normally lasts several days but in extreme cases may last
several years. More frequent outbursts and smaller amplitudes of
light variation are also sometimes cited, but here there is a continuous distribution and no sharp physical division. There are interesting differences in behaviour of different systems.Several systems
have normal outbursts and "super" outbursts. During the latter, such
a system may remain several days at maximum light. A yet unexplained
feature is that so far the interval between supermaxima has been more
regular than that between normal maxima.

3.3.   At minimum light, the brightness is not constant but shows two
types of variations. When studied with the proper equipment, rapid
fluctuations are found; these changes are now called "flickering".
To study this in detail, high speed photometric techniques were necessary. These were first developed by Nather and Warner (1971) and
their colleagues at the McDonald Observatory and provide resolution
in time intervals as small as one second. The nature of the flickering is by no means identical in the different systems, but varies
in its rapidity. In color effects in the cases studied to date, it
is strongest in the ultraviolet as compared to longer wavelengths.

While the precise details of the flickering vary from one system
to another, certain features have been found in all that have been
carefully studied. In addition to the apparently random fluctuations,
observed periodicities have been found which continue for intervals
which are very long compared to the periodicities themselves. A
typical period is of the order of 20 seconds and can remain unchanged

through hours of observation. It is this periodicity or coherence
that differs them from the other rapid oscillations in the light curve
which show no such periodicity. The amplitudes are small, being at
the most of the order of a few hundredths of a magnitude. These seem
to be typical of all cataclysmic variables and are not confined to
the dwarf novae. The physical cause of the periodicity is still not
known. It has been attributed to non-radial pulsations in the white
dwarf component, but hot spots on the surrounding disk have also been
suggested.

In addition to the flickering, the light curves show changes related to the orbital revolution of the components. Sometimes these
appear to be eclipse effects of some sort but other changes are frequently present even when no eclipses are shown. Figure 9.3.1 shows
a schematic light curve of U Geminorum with the flickering effects
removed.

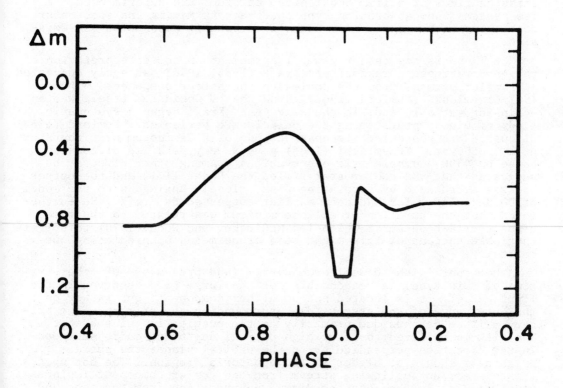

Fig. 9.3.1. Schematic representation of a typical light curve
of U Geminorum. Flickering effects have been removed.

A large increase in brightness beginning at about phase 0.6 P reaches
a maximum at about phase 0.9 P and the light is decreasing at the onset of the eclipse. When the eclipse ends, the brightness is appreciably lower than at the beginning. Indeed the brightness is approx-
mately that indicated by an extrapolation of the light changes immediately preceding first contact. The flickering is strongest during
the bump preceding primary. During the constant phases of primary it
has greatly decreased and is little if any greater than that expected
from normal scintillation effects and photon noise. This cessation
of flickering during eclipses is also shown by UX Ursae Majoris and
RW Trianguli, and appears to be a characteristic of the class.

3.4.   Elvey and Babcock (1943) gave the first general discussion on
the spectra of dwarf novae, and Joy (1943) first suggested their binary
nature on spectrographic evidence. Beginning in the mid 1950's, these
systems were studied increasingly. Joy (1956) noted that SS Cygni
showed two components of spectral types dG5 and sdBe, respectively;
the latter displaying strong emission lines more than 20 Å wide. Joy
suggested that the hot component was surrounded by an expanding shell
or by matter falling towards it in a turbulent manner.

In general, the other systems studied showed at minimum light a
continuous spectrum on which Balmer emission lines, and sometimes
emission lines of helium and ionized calcium, are superimposed. A
few systems show absorption type spectra. At maxima the spectra are
continuous with occasionally weak lines both in absorption and emission.

3.5.   There is now rather general agreement on the interpretation of
the observations. Crawford and Kraft (1955, 1956) had early suggested
a transfer of mass from the cooler to the hotter component. Then in
1961 Krzemiński (Mumford, 1962) found that U Geminorum itself was an
eclipsing variable and, in the same year, Kraft began a systematic
study of binary stars among the calaclysmic variables. Mumford (1964)
discussed his photoelectric observations of U Geminorum near the start
of an outburst. Krzemiński (1965) gave a very detailed discussion
based on 4500 photoelectric observations. Among other effects, he
suggested that the cooler star filled the Roche limit and the hotter
one was surrounded by a rotating disk, closely analogous to the model
of DQ Herculis which had been earlier suggested by Kraft. He further
found that the duration of eclipse minimum was related to the time
since the last outburst which led McCluskey and Wood (1970) to suggest
that observations of this might be a crude means of predicting outbursts.

A hot spot in the U Geminorum system (and presumably of other systems of this type) is responsible for the large hump observed in the
light curve. It is located not on either component, but rather on
the outer edge of the gaseous disk surrounding the primary. The primary eclipse is thus due primarily to the occultation of the spot.
The disk is believed to be of high optical depth and quite extended
in the direction perpendicular to the orbital plane; the surface
brightness should be higher in the equatorial region. The hot spot
exists where an outflowing stream from the cooler component impinges
upon material in the rotating disk; the radiation from the spot and
at least part of that from the disk results from collisional excitation resulting from the kinetic energy of the stream. The emission
lines from the spot blend with double lines from the disk resulting
in complex profiles which vary with phase.

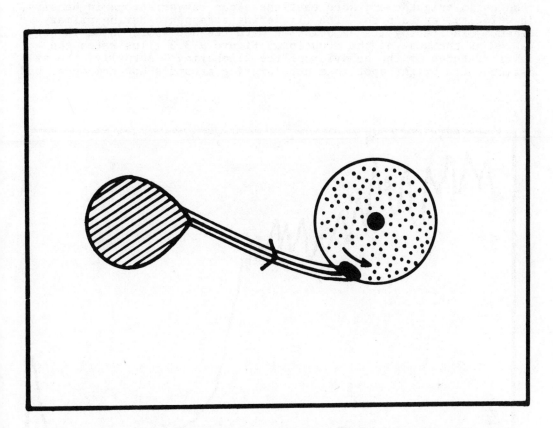

Fig. 9.3.2. Model of a dwarf nova. The size of the disk surrounding the
white dwarf varies with the time ellapsed since the lates outbursts.

Smak suggests that the outburst involves at least two probably related phenomena. Most of the luminosity increase is caused by the
brightening of the central parts of the disk. Shortly before maximum
these are extensive enough to be partially eclipsed and a shallow
broad minimum is then observed with minima occurring earlier than the
spectroscopic conjunction as first reported by Krzemiński. The brightness of the spot itself increases very little if at all; thus its
eclipses are detected during the initial rise and later stages of decline, but not during maxima. Simultaneously, the size of the disk
increases by as much as 30 per cent, as measured by the radius vector
of the spot. During decline, the disk contracts and this continues
at a much slower rate until the next outburst.

A remarkably similar picture, but from quite different evidence was
presented almost simultaneously by Warner and Nather (1971). They
studied the rapid light changes with amplitudes up to 0.3 magnitudes
in times of minutes and of less amounts in times of seconds, and explain this flickering in terms of the impact of a non-homgenous stream

from the secondary upon a disk surrounding the primary. It is the
inhomogeneities in the stream which are responsible for the rapid
changes in brightness. Note that these can be variations in density,
in velocity, or in both. The flickering disappears during primary
minimum, indicating its association with the hot spot, which is suggested as the seat of the eruption. Figure 9.3.3 illustrates the
chief features of the behavior of the flickering. Actually, the existence of a bright spot in a nebular ring around a hot component had

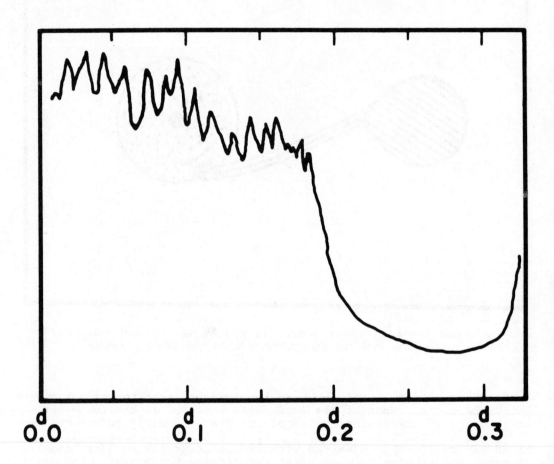

Fig. 9.3.3. Schematic of an eclipse of U Geminorum. The flickering
disappears when the hot spot is eclipsed.

been suggested by Walker and Herbig (1954) as early as 1954 in their
work on UX Ursae Majoris.

    To summarize, the picture that has developed is thus the following.
The primary is assumed to be a white dwarf while the secondary, a
late-type star, fills or very nearly fills its limiting surface. The

system thus is at a relatively late evolutionary stage. The more massive primary has already had most of its mass either tranferred to the secondary or lost to the system.

However, the secondary presumably is now in a stage of expansion and at least some - probably a great deal - of the material is being transferred to within the limiting lobe of the primary and is stored at least for a time, in an accretion disk. We have mentioned the "hot spot" created by impact on this disk of material recently ejected from the secondary. While this picture is, at the time of this writing, rather generally accepted, the masses and other evolutionary aspects are not so clearly established.

3.6. Reliable determination of the masses is made difficult by the fact that only two U Geminorum systems are double lines binaries whose inclination can be determined. EM Cygni is an eclipsing binary and Z Camelopardalis shows light variations which permit its inclination to be estimated to an accuracy of about 5°. For the other systems indirect methods must be used. The most recent discussions are by Robinson (1976) and Ritter (1976).

Their methods use a relation between the orbital period and the mass of the late type star. This makes the usual assumption that this component fills its critical surface and further that it obeys the mass-radius relation for the lower main sequence. When applied to cataclysmic variables in general, no preferred mass is found. Further, it is impossible to distinguish between novae and dwarf novae on the basis of the masses of either component. A range of at least a factor of two exists when intercomparing the masses of either the white dwarf or the late-type components of dwarf novae, while classical novae can be found in which the mass of the white dwarf or that of the later type star is nearly equal to masses found in the dwarf novae.

3.7.   The source of the outburst has also been a matter of debate. When McCluskey and Wood (1970) wrote their review article, arguments could be presented in favor of either the hotter or the cooler component. Now, however, as Smak (1971) and, more recently Warner (1974b) have discussed the photometric evidence strongly indicates that it is the disk which is bright during the outbursts. Lynden-Bell and Pringle (1974) have considered the evolution of viscous disks in general, and included U Geminorum in particular. They consider the outbursts are caused by the "periodic" dumping of mass (10^{-9} $\mathfrak{M}_{\odot}$) onto the white dwarf by its companion; this has in general too much angular momentum to fall directly on the white dwarf and initially forms a circumstellar ring. Bath et al. (1974) have considered the accretion model for outbursts of dwarf novae.

There is as yet no consistent picture of the details. Two models have been developed. One considers unstable nuclear burning on the surface of the blue component. The other invokes unstable mass transfer from the red component. According to this model mass outflow from the secondary falls toward the primary, but, because of its angular momentum, cannot be immediately accreted. It thus forms a disk around the primary. During outbursts the rate of mass transfer is greatly increased. Alternate approaches suggest absorption and reemission by the disk of the radiation from the white dwarf.

3.8.   As noted earlier, the Z Camelopardalis variables are now usually described as a subclass of the dwarf novae, but could be listed, as once they were, as an individual class of eruptive variable. If considered as a separate group, they certainly constitute an exceedingly

small one. The General Catalogue of Variable Stars (1971) lists only
20 members. Here, we will only note a few of their characteristics
and then consider in more detail their prototype, Z Camelopardalis.

The chief distinction from the U Geminorum stars are the intervals
of constant light, sometimes called "standstills" or "plateaus", which
occur in the interval between a maximum and the following minimum.
Normally these occur about one-third of the way down the descending
branch. The frequency of occurrence varies considerably from system
to system. In some they occur relatively frequently while in others
their appearance is much less so. To date with only two exceptions
they have been found only on the descending part of the light curve.
They last for at least a few days but constant intervals of a year or
more have been observed. Almost always, after this interval the drop
to minimum continues; one exception has been observed on one occasion
in Z Camelopardalis itself, and one in TZ Persei when a standstill
ended in each case with a rise to maximum. These are also the two
systems which on one occasion each have been observed to rise from a
minimum to a standstill. Compared to the U Geminorum systems, there
is a tendency to have smaller amplitudes in the outbursts and shorter
intervals between them but there is considerable overlap and these cannot serve to classify a given system. We will now consider Z Camelopardalis itself.

The system is approximately 14th magnitude visually at minimum light
and usually a little fainter than 10th magnitude at maximum. The average interval between eruptions is some 20 days. The standstills occur,
when they do, after maxima when it has dropped about two magnitudes.
Normally these last for a few days, but they have been observed to
continue for up to two years. At times the system shows intervals of
quite irregular fluctuation. One effort by Florkowski (1975) to detect radio emission from the system during minimum light yielded negative results.

While Z Camelopardalis was once reported to show shallow eclipses,
the later evidence indicates that it does not, at least in the conventional sense of stellar eclipses, although there are periodic changes
in the light curve which are almost certainly associated with the orbital motion. The principal one of these is a "shoulder" similar to
that formed in U Geminorum. Outside the major eruptions, the light
curve shows the irregular, rapid fluctuations typical of the dwarf
novae. Kraft et al. (1969) find the maximum of this shoulder is preceded by a (U-B) depression and followed by an interval of (U-B) excess.
Warner and Nather (1972) have reported that at times of conjunction,
as predicted from the spectrographic elements, they have noted slight
dips in the light curve accompanied by a notable reduction in the flickering. This is what would be expected from a grazing eclipse of a
bright spot. This may not always be present because of changes in the
size of the spot and in its distance from the primary component.

An early discussion of the spectrum was given by Elvey and Babcock
(1943) in a general discussion of spectra of U Geminorum stars. Kraft
et al. (1969) gave a more detailed treatment. In common with similar
systems, Z Camelopardalis is semi-detached; the spectral types are G1
and sdBe. Like others in the group, the blue star appears to be a white
dwarf. At minimum, the spectrum is dominated by wide emission lines of
H, HeI, and Ca II. At maximum it is virtually continuous, but feeble
emission lines centered on very wide, shallow, absorption lines of
hydrogen have been reported. Values of $\mathfrak{M} \sin^3 i$ of 0.49 $\mathfrak{M}_\odot$ for the red
star and 0.66$\mathfrak{M}_\odot$ for the blue star were found; the precision, however,

was not considered to be high.   The emission lines are wide and frequently double, but not as wide as those of U Geminorum.

These observations were used by Smak (1970) in his discussion of
spurious effects in the radial velocity curves caused by the blending
of emission lines originating in the spot with those coming from the
disk.   Variations found in the observed amplitude and phase shift were
attributed to variations in the contribution from the spot. $\mathfrak{M}_b/\mathfrak{M}_r$ =
0.96; i.e., nearly equal masses but with the blue star slightly less
massive.   In SS Cygni and RU Pegasi, the blue component also has been
found to be of lower mass than the red.   However, a later study by
Robinson (1976) found masses of $0.90\mathfrak{M}_\odot$ for the red and $1.26\ \mathfrak{M}_\odot$ for the
blue member.   We can only conclude that the question of the masses is
still an open one, although the mass of neither star seems to deviate
by a large factor from the solar mass.

Z Camelopardalis has attracted a good deal of attention in recent
years.   Faulkner (1971) discussed in more detail earlier suggestions
of gravitational radiation losses.   This could reduce the scale size
of the system and thus induce mass transfer at approximately the observed rate.

Robinson (1973a) has made an extensive study of the rapid fluctuations.   Strong flickering is always present and its amplitude is correlated with the position in the eruption cycle.   A periodicity is
found in the light changes during eruptions (and only during eruptions);
these are strictly periodic for hours with periods between 16 and 19
seconds.   On one occasion, the period showed a small, abrupt change
from 16.90 seconds to 16.98 seconds; the change occurred in less than
15 minutes (Warner and Robinson, 1972).   The bright spot on the disk
of gas surrounding the white dwarf contributes at least twenty per
cent of the total luminosity of the system at minimum light.

Later, Robinson, (1973b) studied a series of image tube spectra of
the H_α emission line.   The center of this profile has a complex, triple
peaked structure which does not share the orbital radial velocity variations and apparently comes from an expanding shell which encloses the
entire system.   The shell apparently is not formed by mass ejection
during the eruptions, but rather by continuous mass loss from the disk
surrounding the hotter component.   Mass loss to the shell was computed
to be $2.4 \times 10^{-9}\mathfrak{M}_\odot$ /yr, and this should notably affect the evolution
of the system.

Thus, while the picture of Z Camelopardalis and other members of
this group seems reasonably clear in most major details, many puzzles
still remain.   Perhaps the most serious question is the physical cause
for the long standstills on the way to light minima.

3.9.   Our knowledge of the evolution of cataclysmic variables is at
present highly tentative.   Sahade (1959) and Kraft (1962, 1963) early
suggested their progenitors could be the W Ursae Majoris systems.   A
certain amount of support was given by observed space distribution and
motions, e.g., Kraft (1966), Warner (1974a).   Warner further suggests
that the cataclysmic variables may evolve into one kind of Type I supernova; the mechanism could be the transfer of He to the white dwarf primary, pushing it over the Chandrasekhar limit.   Ritter (1976), however,
has explained the existence of these systems as results of normal stellar evolution through mass exchange which strips the whole envelope of
the developing red giant to reveal the white dwarf.   If the mass exchange starts before central helium burning, the resulting star would
have a mass less than 0.45 $\mathfrak{M}_\odot$; if it began after the central helium

ignited, the final mass would be greater than this limit. If these
ideas are adopted, the W Ursae Majoris systems cannot be the progenitors of the U Geminorum stars (or any other cataclysmic variables)
since a high initial angular momentum is needed for a "massive" white
dwarf to be formed. The entire picture is far from clear.

Finally, we note that the model presented is obviously a highly
simplified version of a complex situation. The structure of the envelope must be extremely complicated both because of the inhomogeneity
of the impacting stream and the rotation of the envelope. This idea
is strengthened by observed difference in the light curves of different
epochs. The discovery of intense "ultrasoft" X-ray emission at the
time of one light maximum (cf. IAU circular No. 3125) is so recent that
its full implications have not yet been discussed. In fact, it is surprising that a relatively simple model fits so many of the observed
data.

4.   NOVAE

One of the most important landmarks in our knowledge of close binary systems and their evolution was the discovery by Merle Walker (1954)
that Nova Herculis 1936 (=DQ Herculis is an eclipsing variable with a
period of 4^h39^m. Earlier the recurrent nova T Coronae Borealis (Sanford, 1949) and the "flaring" variable AE Aquarii (Joy, 1954a, 1954b)
were found to be spectroscopic binaries. The question then arose as
to whether all novae and nova-like stars - and, in fact, all catalys-
mic variables - are close binary systems and as to whether the eruptions
were somehow triggered by the binary nature of the objects (Struve,
1955). An extended investigation by Kraft and others (Kraft, 1962,
1964; Kraft et al., 1962) on the velocity behavior of a number of U
Geminorum stars and old novae did suggest that they are all indeed binaries, and further work by several other astronomers confirmed this
conclusion, which is now in the category of a well established fact.

As Faulkner (1974) states, "although one can attempt to contrast
and distinguish between the four types: classical novae, recurrent
novae, dwarf novae and nova-like variables, the distressing fact is
that in many of their properties there is considerable overlap". We
quote in Table 9.4.1. the outburst data that characterize them, as
compiled by Robinson (1976).

It seems, therefore, more appropriate to discuss all cataclysmic
variables together, and we should mention three recent review papers
that do so, namely, those of Smak (1971), Faulkner (1974) and Robinson
(1976). The group of cataclysmic variables, as characterized by the
four types of objects mentioned, have very short binary periods, $1.5\mathfrak{M}_\odot$
hours to 18.5 hours, except for T Coronae Borealis, which has a period
of nearly 2/3 of a year.

In planning this book we have, however, chosen to deal with the
dwarf novae and with the "classical" and recurrent novae in separate
sections, just to emphasize each group individually. Moreover, we
are discussing in our next section the symbiotic stars, which also
undergo nova-like outbursts and are probably all binaries.

Smak (1971) lists the symbiotic binaries as objects that are related to the eruptive or cataclysmic variables. In both cases we are
dealing with late stages in the evolution of close binaries and they

Table 9.4.1.

OUTBURST CHARACTERISTICS OF CATACLYSMIC VARIABLES

Objects	Light Amplitude (magnitudes)	Energy (ergs)	Recurrence Time (years)
Novae	9 – > 14	10^{44}–10^{45} or more	Only one outburst
Recurrent novae	7 – 9	10^{43}–10^{44}	10–100
Dwarf novae			
a) U Geminorum stars	2 – 6	10^{38}–10^{39}	0.04–1.4
b) Z Camelopardalis stars	2 – 5	10^{38}–10^{39}	0.03–0.15
Nova-like variables	–	–	No outbursts

should be considered under the same umbrella. However, the orbital
periods that characterize the symbiotic binaries are rather long -
of a few years, at most - and suggest that their evolutionary history
may be quite different from those of the dwarf novae or the "classical" novae.

How can we describe a binary system which is a nova? Their spectra
display double lines in a very few cases like T Coronae Borealis, GK
Persei, V1017 Sagittarii. In these three objects one component is an
underluminous blue star while the companion is a late type giant, M3III
in T Coronae Borealis, G5III in V1017 Sagittari and G2IVp as an average,
in GK Persei (Kraft, 1964). In the rest of the cases, which have orbital periods of hours, the evidence is - at least through the presence
of emission lines - for only an underluminuous hot component, the companions being perhaps late type dwarfs, as is true in the case of the
dwarf novae. The blue star is probably in most instances, if not always, a white dwarf, and such a conclusion is the result of two observational facts.

The first observational fact is related to another remarkable discovery that resulted from Walker's photoelectric observations of DQ
Herculis. The light curve, outside of eclipse, is characterized by
oscillations that have a periodicity of 71^s and an amplitude of 0.04
magnitude. The period of these fluctuations decreases at the rate of
2.7×10^{-5} sec yr-1 (Herbst et al., 1974); furthermore, Warner et al.
(1972) have shown that, through eclipse, the oscillations shift $+ 360^\circ$
in phase. This light behavior can be understood in terms of non-radial
pulsations of a magnetic white dwarf, its core rotating like a rigid-
body with a period of 71^s. In this model most of the visible light
emitted by the system results from the reemission of absorbed ultraviolet radiation by matter located near the white dwarf (Katz, 1975).

The second argument for white dwarfs being one of the components in novae, is provided by the fact that the spectra of WZ Sagittae - a recurrent nova which is an ultrashort period eclipsing binary -, of Nova Lacertae 1910 (=DI Lacertae) and of UX Ursae Majoris - a nova-like variable which has a light behavior and a periodicity very similar to those of DQ Herculis - display the broad and shallow absorption features that are characteristic of white dwarfs.

The binary system model, in the case of the novae, was first suggested by Kraft (1959, 1963, 1964) as a generalization of his interpretation of T Coronae Borealis (Kraft, 1958), DQ Herculis (Kraft, 1959) and the U Geminorum stars (Kraft, 1962). The red component overflows the critical equipotential lobe and gaseous material streams towards the white dwarf companion forming an accretion disk around this star, which appears to be the seat of the eruptions. The emission lines that characterize the spectrum of the binary would originate in the disk.

We have mentioned two recurrent novae, namely, T Coronae Borealis and WZ Sagittae. While T Coronae Borealis is also a symbiotic object, WZ Sagittae is an ultrashort period binary that behaves similarly to a dwarf novae. Actually, Krzemiński and Smak (1971) have proposed a model for WZ Sagittae by adding a "hot spot" in the disk that surrounds the white dwarf component. This model, which assumes that the white dwarf is the more massive member of the pair, appears to account for the light and spectral behavior of the system. The spectrum of WZ Sagittae can be described as displaying (Krzemiński and Smak, 1971)

a) a hot continuum;

b) broad, stationary H absorption

c) double H emissions superimposed upon the cores of the absorptions, separated by 1430 km sec^{-1}, according to Greenstein (1975);

d) a so-called S-wave emission component yielding a velocity curve, shifted by more than $90°$ with respect to the light curve, with an amplitude between 650 and 850 km sec^{-1};

e) a weak Ca II-K absorption of probably interstellar origin.

The eclipsing light curve occasionally becomes similar to that of U Geminorum (see section 3 of this Chapter).

In the model for WZ Sagittae the outer parts of the disk have a rotational velocity of 715 km sec^{-1} and the hot spot is responsible for the "S-wave emission component". Observations by Boksenberg and Greenstein (cf. Greenstein, 1975) with the University College, London, Image Photo Counting System may throw further light on our understanding of the object.

The fact that T Coronae Borealis and WZ Sagittae are recurrent novae suggests that we have the configurations of the symbiotic objects or of the dwarf novae depending on the volume available for the evolutionary expansion of the presently red component before mass streaming sets in. On the other hand, the observations strongly suggest that the common feature is the presence of a condensed, white dwarf companion, which is the seat of the outbursts.

What can we say about the outburst mechanism that is relevant in classical and - presumably - in recurrent novae? As Robinson (1976) puts it "there are several sophisticated theories of the nova eruptions

but they meet with only partial success when confronted with observations. The only theories of the eruption that have successfully
produced this much energy {10^{45} ergs or more} are those in which the
energy source is a thermonuclear runaway in the envelope of the white
dwarf...but they cannot reproduce the gross morphology of the light
curves of classical novae (Starrfield et al.,1974)." This, however,
may only be due to the neglect of the accretion disc and the circumstellar material that surrounds the binary (Starrfield et al, 1975).

The fact that we are certain that all novae are binaries does not
necessarily mean that we have a fairly good knowledge on their masses.
The contrary is true. Moreover, it has not even been possible to
ascertain whether the mass of the red component might be always smaller or larger than that of the white dwarf. To illustrate our present knowledge on the masses we quote in Table 9.4.2. the presently
available "direct" values.

Table 9.4.2.

MASSES OF COMPONENTS IN "CLASSICAL"

AND RECURRENT NOVAE

Star	$\mathfrak{M}_{red\ star}$ ($\mathfrak{M}_{\odot}$)	$\mathfrak{M}_{blue\ star}$ ($\mathfrak{M}_{\odot}$)	Notes
T Coronae Borealis	≥ 2.9	≥ 2.1	1
GK Persei	≥ 0.56	≥ 1.29	2
DQ Herculis	~ 0.20	~ 0.12	2

Notes to Table 9.4.2:

1. Kraft, R. P., Astrophys. J. 127, 625 (1958). We are quoting
    minimum values.
2. Kraft, R. P., Astrophys. J. 139, 408 (1964).

In novae as in all cataclysmic variables we must be dealing with
binaries undergoing a second episode of mass loss.

We should not close this discussion without mentioning two points. In the first place, the light of DQ Herculis has been found to be polarized and that the polarization is variable (Nather et al, 1974; Swedlund et al., 1974; Kemp et al., 1974). It is, however, not yet clear whether the periodicity involved is the same as that of the light oscillations or twice as long.

Secondly, we have the case of Nova Cygni 1975 (=V1500 Cygni). For the first time the techniques of high-speed photometry were applied to a nova very shortly after its outburst. The observations suggested a period of light variation of the order of 0.14 days or perhaps 0.28 days, which gradually decreased from the middle of September, 1975 to the end of June, 1976 (cf. Semeniuk et al., 1976), that is, in a ten month interval that starts from about two weeks after the nova maximum of light. Rapid variations in spectral line intensities have been also detected (Campbell, 1976; Young et al., 1977).

V1500 Cygni is the first nova for which observations with different techniques carried out almost since the outburst are available. They will provide us with an invaluable opportunity to understand better the processes involved in the nova phenomenon.

Readers interested in learning about novae in general are referred to review papers by McLaughlin (1960), Greenstein (1960) and Mumford (1967) and to volumes like the one that contains the proceedings of the recent colloquium on novae and related stars (Friedjung, 1977).

5.   SYMBIOTIC STARS

The symbiotic stars form a group of objects characterized by spectra which display a combination of low-temperature absorption lines and emission features that require high excitation conditions. The prototype has been always considered to be Z Andromedae because it is the first star of the kind ever investigated in detail.

In 1941 the objects in the group were called "symbiotic stars" by Paul W. Merrill, who used the expression "combination spectra" to describe their spectral characteristics. The two terms have been in use ever since.

The symbiotic stars are quasiperiodic variables both in light and in the spectral features, with the increase in brightness being steeper than the decline. The absorption spectrum that is displayed corresponds to that of a normal late type (commonly M) giant, which is prominent when the star is faint. When the star brightens, an early type shell spectrum develops; this dominates the photographic region and covers the late-type spectral lines. As the star declines in brightness, the shell spectrum weakens and emission lines, of progressively increasing excitation, and forbidden transitions develop.

The general behavior suggests that we are dealing with objects that undergo quasiperiodic nova-like outbursts, which lead to spectral changes similar to those of a slow nova. These changes superimpose upon a normal late-type giant spectrum which apparently does not undergo noticeable variations in itself.

In addition, we should stress the fact pointed out by Boyarchuk (1969) that "irregular variations in brightness is one of the most characteristic features of symbiotic stars".

Several hypotheses have been advanced to explain the symbiotic objects. These hypotheses can be grouped into two catagories, namely, the ones that postulate that we are dealing with a single object and those that assume that we are dealing with binary systems. The latter hypothesis, first suggested by Hogg (1934), is the most generally accepted; it finds support on three main counts, namely,

a) it provides the simplest explanation;

b) a number of symbiotic stars have been shown to be binary systems;

c) the two objects that would form such binaries provide the combination that appears to lead to eruptive activity.

We are not going to give a detailed account of everything that is known about the symbiotic stars. We shall confine ourselves to mention the cases that have been found to be physical pairs and refer the reader to the several review papers that have been written on the subject, namely, those of Payne-Gaposchkin (1957), Merrill (1958), Sahade (1960a, 1965), Boyarchuk (1969), Swings (1970) and Sahade (1976b).

The cases where binary nature has been already established are listed in Table 9.5.1 Boyarchuk (1969) mentions three additional objects that appear to display periodic variations in radial velocities, namely, BF Cygni, RW Hydrae and R Aquarii, and Payne-Gaposchkin (1977) quotes orbital periods for Z Andromedae, CI Cygni and SY Muscae.

We are, therefore, dealing with a group of binaries formed by a giant late type star and an early type subdwarf. This statement finds additional support in the fact that the visual absolute magnitudes of the symbiotic objects is -3 or -4 and, therefore, if the late type component is a normal giant, then the hot companion should be a subluminous star located below the main sequence. The periods of the orbital motion appear to be of very few years at most.

In general it is accepted that these symbiotic objects are interacting binaries and that the giant component probably fills the equipotential lobe. The subluminous star is the component with which the nova-like behavior is associated.

If this is so, we find that the characteristics that are shared by the objects that undergo eruptive variability are

a) binary nature;

b) a hot subdwarf component;

c) a component filling the equipotential lobe.

and in all cases we must be dealing with a second episode of mass loss. Plavec (1973) has suggested that at least some symbiotic stars may be examples of case C of binary evolution; the first process of mass outflow must have, then, been of mode B to lead the present blue star to its current location below the main sequence. The galactic distribution of the symbiotic stars and their radial velocities suggest that they are population II objects (cf. Thackeray, 1973).

Let us consider now our knowledge of the masses. The velocity curve of the late type component is usually well determined. However, because

Table 9.5.1.

SYMBIOTIC BINARIES

Star	Period (days)	Spectrum	$f(\mathfrak{M})$	Remarks	Notes
17 Leporis	260	M1 III + B9	0.24		1
AX Monocerotis	232	gK + B3nn	3.0		2
T Coronae Borealis	230	M3 III + sdBe		recurrent nova	3
AR Pavonis	605	M3 III + sd		eclipsing	4
AG Pegasi	820	M3 III + ...	0.014		5,6

Notes to Table 9.5.1.:

1. Cowley, A. P., Astrophys. J. 147, 609 (1967).
2. Cowley, A. P., Astrophys. J. 139, 817 (1964).
3. Kraft, R. P., Astrophys. J. 127, 625 (1958): the derived minimum masses are $2.9\mathfrak{M}_{\odot}$ for the M3 III component and $2.1\mathfrak{M}_{\odot}$ for the subdwarf companion.
4. Thackeray, A. D. and Hutchings, J. B., Mon. Not. R. astron. Soc. 167, 319 (1974).
5. Cowley, A. and Stencel, R., Astrophys. J. 184, 687 (1973).
6. Hutchings, J. B., Cowley, A. P. and Redman, R. O, Astrophys. J. 201, 204 (1975).

of the nature of the system and its spectral behavior, it is difficult to secure a reliable velocity curve of the early type star or even be certain as to which spectral lines could purely yield its orbital motion. Thus the derivation of the masses that are assigned to the components is made only after some interpretation of the data and/or the adoption of certain assumptions. Perhaps the best case, as far as the mass determination is concerned, is that of T Coronae Borealis. In this object the velocity curve of the blue component was derived from the measurement of the Hβ emission after reconstructing its profile. This velocity curve combined with the one yielded by the late type component led to the minimum masses quoted in Table 9.5.1. and, therefore, to the conclusion that the blue subdrawf has a somewhat smaller mass than the M3 giant (Kraft, 1958). The analyses of 17 Leporis, AX Monocerotis and AR Pavonis suggested that the mass situation is just the reverse, namely, that, in each case, the more massive component is the blue star. However, how much we can rely on

the masses inferred for these systems remains as a problem which a-
rises because of the difficulty of the systems. As a result, in Table
9.5.1 we are listing only the values of the mass-functions as derived
from the velocity curves of the late type components, whenever available.

AG Pegasi deserved special mention. The investigation of this system by Hutchings et al. (1975) has led to the conclusion that the
mass of the late type giant is larger than that of the companion and,
furthermore, that the gaseous stream in the system goes from the blue
component towards the M3 giant. The latter conclusion is at variance
with the suggestions from the other symbiotic binaries and with what
we would expect if the blue component is the seat of the nova-like
outbursts. To have a clear model of the symbiotic binaries we certainly need further work.

Among the symbiotic stars, even in the small sample of Table 9.5.1,
we seem to be dealing with actually somewhat different objects -- T
Coronae Borealis is, for instance, a recurrent nova, as is RS Ophiuchi
which is also a symbiotic star -- to the extent of the character of
the outbursts, even though the group seems to be homogeneous insofar
as the components of the pairs is concerned. This may be connected
with how far the blue component is advanced in its evolution.

Webbink (1975, quoted in Bath, 1977) has suggested that "symbiotic
variables consist of main sequence stars accreting at super-critical
accretion rates from giant companions". Furthermore, Bath (1977) has
worked out a binary model for the symbiotic objects, which he compared
with the behavior of Z Andromedae. This model assumes that the hot
star, through a disk, accretes matter shed by the late type companion
at rates "close to (and occasionally in excess of) the Eddington limit.

In the symbiotic stars we obtain information about the structure
and evolution of the outer envelope that is produced by the outbursts.
In AG Pegasi, for instance, we observe absorption lines that suggest
the effect of diluted radiation (HeI λ 3888) and imply electron densities of the order of 10^{11} cm^{-3}, and also forbidden emissions that
suggest electron densities of about 10^7 cm^{-3}.

There are objects that seem to be intermediate between the symbiotic stars and the planetary nebulae -- and it has been suggested (cf.
Sahade, 1976b) that at least some of the symbiotic objects may develop into planetary nebulae. Sahade further suggested that to investigate such a possibility implies "to investigate the nature of the
central stars" of the planetaries. We shall discuss in the next
section of this Chapter what we know about some of them at the present time.

6.   CENTRAL STARS OF PLANETARY NEBULAE

In a recent paper on the spectra of the central stars of planetary nebulae, Aller (1977) lists six different types of spectra that
characterize these objects. A number of them are Wolf-Rayet, Of or
O stars but there are some which show only a continuous spectrum and
others that have low-excitation spectra. The latter pose the problem
of tracing the source of excitation of the nebulae.

Earlier the tendency was to think that the stars that do not display high-temperature features could not be physically connected with planetary nebulae, and that they were probably foreground objects. Of course, the possibility existed that these low-temperature central stars were physically associated with early type subluminous companions, but this idea began to open its ways through only about a decade ago. Thus, the existence of low-temperature objects among central stars of planetary nebulae is only a recognized fact as of late.

The first case of a physical pair associated with a planetary nebulae was that of NGC 246. Its central star is an early type (O) subdwarf which has a visual companion of spectral type G8V or K0V, 4" away (Minkowski, 1965).

This discovery was followed by the first analysis of the case of a central star with spectrum later than O, namely, BD + 30°623 (A0 III), which is associated with NGC 1514. Kohoutek (1957, 1968) and Kohoutek and Hekela (1967) reached the conclusion that in this case we must be dealing with a close pair, where the companion to the A0 III star should be an O-type subdwarf. Variations in radial velocity have been reported by Kohoutek (1968) and by Mannano et al. (1968) and a plausible semiamplitude range appears to be of the order of 25 km sec^{-1} (Greenstein, 1972).

In this avenue of research that has been recently opened there are two additional "firsts" that we should mention. One is the rediscovery of UU Sagittae, which coincides with the central star of Abell 63 (Bond, 1976), as an eclipsing variable (Miller et al., 1976). The period of this binary which has an O-type spectrum is 11^h09^m6; "the eclipse lasts 70 min with a flat-bottomed minimum of about 16 min". The drop in light during the eclipse is 4.3 magnitudes. According to the Card Catalogue of the University of Florida the actual discovery of UU Sagittae as an eclipsing variable is due to Dorrit Hoffleit (1932).

The second important development refers to the result obtained by Acker (1976) in her investigation on the nuclei of planetary nebulae. From her work it appears that FG Sagittae, the central star of He1-5, and BD + 30°623 are single-lined spectroscopic binaries, while BD + 66°1066, associated with NGC 6543, seems to be a double-lines spectroscopic pair. In the latter case, the radial velocities from one of the components being derived from the HeII λ 4686 Å emission line.

Acker's findings were followed by Méndez and Niemelä's (1977) study of CPD -26°389, the subdwarf O central star of NGC 1360. This object was found to be a single-lined spectroscopic binary with a total mass probably below one solar mass and the O component between 0.04 and 0.3 solar masses.

A few candidates are presently being investigated for their binary nature. Table 9.6.1 pictures the present situation, and we have included the known visual binaries for the sake of completeness.

Several groups of objects have been suggested as having their evolution going through the planetary nebula stage, and there are statements that predict that the next step is that of a white dwarf. With the discovery that at least some central stars of planetary nebulae are binary systems, we are entering an important era in the study of planetary nebulae. Up to the present time our knowledge of the masses of the central stars of planetary nebulae was based on theoretical considerations, and the generally accepted value has been of the

Table 9.6.1.

BINARY SYSTEMS AMONG
CENTRAL STARS OF PLANETARY NEBULAE

Planetary Nebula	Central Star	Spectrum	p (days)	$f(\mathfrak{M})$	Mass-Ratio	Notes
Visual Binaries						
NGC 246	BD $-12°133$	sdO				1
NGC 6853		O				
Spectroscopic Binaries						
He 1-5	FG Sagittae	G*	18.7	0.24		2
NGC 1360	CPD $-26°389$	sdO	∿8	0.19		
NGC 1514	BD $+30°623$	AOIII(+sdO)	0.41	0.004		2
NGC 6543	BD $+66°1066$	WR+O	0.06	2×10^{-4}	0.8	2
Eclipsing Binaries						
Abell 63	UU Sagittae	O+...	0.46			
Possible Spectroscopic Binaries						
NGC 2346		A				3,5
Possible Candidates						
NGC 3132	HD 87892	AO				4
He 2-36		A				5
Abell 14		F				6
He 2-467		G+...e				7
M 1-2		sgG2				8

*In the interval 1955-1972 the spectrum changed from sgB4 to F6Ia (cf. Christy-Sackmann and Despain, 1974).

Notes to Table 9.6.1:

1. Cudworth, K. M., Pub. Astr. Soc. Pacific 89, 139 (1977).
2. Acker, A., Thèse, Pub. Obs. Astr. Strasbourg 4, fasc. 1 (1976).
3. Kohoutek, L. and Senkbeil, G. Les nébuleuses planétaires, Institut d'Astrophys. Liège, p. 485 (1973).
4. Méndez, R. H., Astrophys. J. 199, 411 (1975).
5. Méndez, R. H., private communication (1975).
6. Abell, G. O., Astrophys. J. 144, 259 (1966).
7. Lutz, J. H., Lutz, T. E., Kaler, J. B., Osterbrock, D. E. and Gregory, S. A., Astrophys. J. 203, 481 (1976).
8. O'Dell, C. R., Astrophys. J. 145, 487 (1966).

order of the mass of the Sun (cf. Smith, 1968). We can now attempt
to obtain direct determinations of the masses involved and reach a
better understanding of the evolutionary paths that lead to such exciting objects.

7. ATMOSPHERIC ECLIPSES

We are going to deal in this section with binary systems that are
eclipsing variables and are characterized by one component being a
late type supergiant with a greatly extended atmospheres. When this
component passes in front of the companion, normally an early type
main sequence star, we then observe the effects of an "atmospheric
eclipse"; this occurs at the phases when the radiation of the early
type component shines through the atmosphere of the supergiant star.

Systems formed by an early type main sequence object and a late
type supergiant are known as the VV Cephei systems, a group first
distinguished by Bidelman (1954) and reviewed recently by Cowley
(1969).

Four systems have been more thoroughly studied for atmospheric
eclipses, namely, ζ Aurigae, 31 Cygni, 32 Cygni and VV Cephei itself.
They have been sometimes called ζ Aurigae stars, because there are
differences in the spectroscopic behavior of the first three stars
relative to those that are considered as VV Cephei stars proper.

AZ Cassiopeiae (Meńdez, Münch and Sahade, 1975), a VV Cephei object, has been found to show in its spectrum the effect of atmospheric
eclipses. Another system that appears to display photometric effects
similar to those that are typical of atmospheric eclipses in AL Velorum (Wood and Richardson, 1964; Wood and Austin, 1978); however,
since its spectrum has been reported to be that of a normal giant,
the object may not belong to the group. The system of ε Aurigae has
occasionally been listed in the same group as the stars mentioned
but the situation in this case is more complex and the system warrants
a separate discussion which is given in section 5 of Chapter 10.

In the broadest sense, Wolf-Rayet binaries could be considered as
giving rise to "atmospheric" eclipses. We have mentioned the light
curve characteristics of V444 Cygni in section 2 of this Chapter.

Indeed, when we consider the relative rarity of very massive systems, and the relatively short time they supposedly spend in this
evolutionary state, it is surprizingly fortunate that we have a number of them near enough to the Sun for detailed analysis. There now
seems little doubt that in all these systems, the eclipses are chiefly
atmospheric rather than bodily. That is, the limb of the late type
star does not act like a sharp edge cutting off at any instant the
same fraction of the B-star's radiation in all wavelengths, but the
eclipse shows "atmospheric" or extinction phenomena with the fractional
light loss at any time depending on the wavelength band being observed. The analogy of a planet setting in a smokey atmosphere and dis-
appearing before it reaches the horizon has sometimes been evoked.

An excellent summary of the work, particularly the spectroscopy,
until 1960 has been given by Wilson (1960). At that time only ζ
Aurigae and 31 Cygni could be discussed in great detail, although 32

Cygni and VV Cephei were also included. A later article by Wright
(1970) gave an extremely thorough discussion of what he termed the
"Zeta Aurigae Stars", again emphasizing especially the spectrographic
material. This paper includes reproductions of the spectra and sample
copies of intensity tracings. Absolute dimensions were computed from
spectral types and luminosities and were compared with values derived
from the eclipse data. Wright (1973) has also given an updated summary of this material.

All four of these systems show certain features in common. When
observed spectrographically, the K-line of Ca II becomes noticeably
strengthened long before the eclipses can be detected photometrically
and remains so long after the hot star is "out of eclipse". The H
line of Ca II apparently behaves in much the same manner, but as
Wright points out, the Hε line is only two angstroms away and this
naturally creates difficulties in interpretation. The conventional
interpretation of the change is the line absorption of the hot star's
light by the tenuous outer atmosphere of the cooler. Detailed study
shows observable differences from eclipse to eclipse and even between
ingress and egress of the same eclipse. A careful comparison of the
K-components of ζ Aurigae, 31 Cygni, and 32 Cygni, made from spectra
taken during totality, has been published by Faraggiana and Hack
(1966). Photometrically, all four show evidence of "atmospheric"
eclipses, as will be discussed in more detail later.

Here, we will summarize the basic information for each system and
describe briefly some of the methods that have been developed to
interpret the photometric observations during the partial phases.

ζ Aurigae. The history of the study of this system is a long and
interesting one. Even in the 1890 Draper Catalogue evidences of peculiarity were found classified as M?, G?, A, and F; on one plate re-
marks were "bright lines" and "spectrum peculiar". Miss Maury (1897)
described the spectrum as composite. Harper (1924) first observed a
single line velocity curve and computed spectrographic orbital elements.
The first suggestion of an eclipse was made by Bottlinger (1926) based
on Harper's statement that the B-spectrum was not present seven days before the inferior conjunction of the K-component. The first eclipse
actually observed photometrically was in 1931-32; Guthnick and Schneller (1932) discussed this extensively. The system was observed
frequently during the interval until the next favorable eclipse in
1939-40 without adding a really significant increase to our understanding of it. The orbital period of 2 2/3 years means that the
system is favorably located for observation only every third eclipse.

The 1939-40 eclipse was observed photoelectrically at several
places, but unfortunately there was no coordination. Note that the
length of the partial phases is such that no one observer can cover
them completely; results of various observers using different instruments must somehow be brought together for a joint discussion. The
observed light loss at any instant is extremely sensitive to the
effective wavelength of observation so the combination of different
observers using different telescope-photometer combinations, made
the treatment of the results difficult. For example, in discussing
the photometric observations, one writer treated them as results of
occultation by an opaque limb of the K-star; another found that they
could be explained equally well by such an occultation or by an atmospheric eclipse; yet a third, by the inclusion of data considered
of dubious value by the first two, postulated the existence of a

huge chromospheric cloud, followed by a "transparent region" or tun-
nel. These results were reviewed in some detail by Roach and Wood
(1952). The conclusion seemed clear that non-filtered (i.e., one
color) photometry - the only kind in existence at that time for
photomelectric measures - could not solve the question of the nature
of the eclipse mechanism.

In the 1947-48 eclipse, a relatively few observations in the ultraviolet, in combination with those of several observers at about λ 4500
Å rather clearly indicated that the eclipse was "atmospheric". The
indications were:

(1) The eclipse started at least 24 hours earlier in the ultraviolet than in the λ 4500 region.

(2) At any given time, the fraction of light lost in the ultraviolet exceeded that of the longer wavelength region. This was predicted by the extinction mechanism proposed by Roach and Wood, but
could not be explained by "normal" eclipses where, except for small
second order effects caused by limb darkening or gravity brightening,
the fractional light loss should be the same, regardless of the wavelength of observations.

Supporting lines of evidence, although somewhat weaker were:

(3) The observed difference in slopes of log τ (optical depth)
versus time could be explained by the curve of growth effect of the
spectral lines.

(4) The atmospheric opacity gradient deduced from the photometric
studies agree with that for the density gradient found from entirely
independent spectrographic studies.

Finally:

(5) The value of the total extinction of the light of the B-star
by the absorption lines in the K-star's atmosphere is effectively
equal to that observed photometrically. This implies no continuous
extinction, and is the weakest part of the treatment. The light loss
observed photometrically is determined with precision, but although
Mulders had earlier shown that the cumulative effect of the Fraun-
hofer lines in the solar spectrum was a significant general absorption, the quantitative application to the K-component of ζ Aurigae
involved a somewhat dangerous extrapolation.

Various spectrographic studies were made of this eclipse (e.g.,
McLaughlin, 1948; Mckellar and Petrie, 1952; Wilson and Abt, 1954)
and several interesting features emerged. In fact, these laid the
basis for much of our present interpretation. Comparison with earlier
eclipses (1934, 1937, 1939-40, 1942) showed an approximately uniform
extent of atmosphere except for the 1937 ingress when it was much
greater than the others. All eclipses showed an asymmetry with absorption during ingress being much stronger than during egress. The
observations suggested clouds in the upper layers of the atmosphere.
It became clear that the chromospheric structure was complex and
must occur as small (10^3 km), rather dense condensations. The phenomena shown at egress at the 1947-48 eclipse were quite similar to
both ingress and egrees in 1938-39, but at ingress of the later eclipse
lines of all types showed a singnificantly lower gradient especially
at greater chromspheric heights. Other parameters (e.g., excitation
temperature, turbulent velocity, degree of ionization) were, however,
quite similar.

Several series of observations, both photometric and spectrographic, were made at the 1955-56 eclipse. Useful data was gathered, but there were no major changes in our understanding of the system. One unusual feature was that at egress the chromosphere was more extended than at preceeding eclipses; the ingress was "normal", although in this system, changes in the extent at ingress from eclipse to eclipse had been common while the egress had tended to remain the same. For the 1963-64 eclipse, an organized collaborative program, sponsored by Commission 42 of the International Astronomical Union, was coordinated by K. Glydenkerne. A serious effort to obtain photometric observations which could be combined easily and accurately was made by the distribution to the observers of identical sets of relatively narrow band interference filters, (Gylderkerne and Johansen, 1970).

Various efforts have been made to interpret the spectra. In a series of papers, Saito (1965, 1970, 1973) has considered explanations for the chromospheric turbulence deduced from the spectroscopy. First, a treatment of shock-pulse propagation was applied to the chromspheres of ζ Aurigae and 31 Cygni; in this model these are built up from acoustic waves generated in the hydrogen connection zone. The observed chromospheric condensations agree with those predicted. Further, on the assumption that the Ca II emission lines originate in the upper chromosphere, the maximum gas velocities of the shock pulses are found to be the same as the turbulent velocities deduced from the width of the lines.

Next was attempted an interpretation of the chromospheric radial velocities observed in the K-component of ζ Aurigae. It is well known that the chromospheric lines are almost always displaced towards the red (with respect to the orbital velocity) on the ingress side and towards violet on egress, and that the deviations from the orbital velocity are larger following ingress than prior to egress. Further, the velocity deviations tend to rise to a maximum after egress and then to decrease. These findings of course are not explained by simple rotation. Wilson had suggested a chromospheric equatorial current with sizable concentrations of matter moving with different velocities. Saito's study concludes that the velocity field in the "intrinsic chromosphere" is composed of a varying velocity field of gas clouds moving with different velocities; this overlaps the mean velocity field which consists of an expansion of gas away from the star. The mass ejection is $10^{-8} \, \mathfrak{M}_{\odot} \, yr^{-1}$. On the egress of the chromosphere, the HI-HII boundary is an ionization front; therefore, the HI gas is accelerated and forms a condensed HI region. Its range in chromospheric height depends on the rate of mass ejection.

Finally, observations of the Ca I λ 6572 line taken after fourth contact in the 1971-72 eclipse, show evidence of a circumstellar cloud indicating a mass loss from the K-component of about $10^{-7} \mathfrak{M}_{\odot} \, yr^{-1}$. This is ten times the mass loss earlier derived from the observed radial velocity differences on the ingress and egress sides.

Saito and Kawabata (1976) found this same line at all phases observed (i.e., around fourth contact and near mid-eclipse) in the 1974 eclipse, although the intensity and velocity deviation from the principal line were considerably less than in 1971-72. Again, they attribute these to an expanding cloud of gas, and the differences at the two eclipses as due to a change in the chromospheric activity and/or the mass loss rate. Kitamura (1967) found general agreement between the observed profile of the K-line and one computed assuming a turbulent velocity of 15-19 km sec^{-1}

The photometric observations also indicated that the system was complex. The intrinsic variability of the K-component, shown by variations in brightness during totality, has long been established One puzzling feature of the photometry is the apparent change in duration of the eclipse.

Roach and Wood had summarized the earlier information which could be explained by a pulsation or by a secular change caused by the expansion of the K-star. Interpretation was difficult because only unfiltered observations were available at the time. Later Kiyokawa (1967) discussed the results of five well-observed eclipses from 1935 to 1963-64 and from increases in duration of totality, suggested a gradual expansion of the K-component. However, Kitamura (1974) re discussed the results of the 1963-64 eclipse using only narrow band observations, including some by C. K. Gordon which had only recently been published. He found a duration smaller than that of earlier eclipses and concluded that the idea of a gradual increase was not tenable. The question must await careful observations of future eclipses.

The reliability of the orbital elements have been briefly discussed by Batten (1967) and by Koch et al. (1970). From tracings, Lee and Wright (1960) found a value for K_2 of about 28 km sec^{-1} which gives $\mathfrak{m}_1 \sin^3 i = 6.2\,\mathfrak{M}_\odot$ and $\mathfrak{m}_2 \sin^3 i = 3.2\,\mathfrak{M}_\odot$. Popper (1961), however, finds values of $8.3\,\mathfrak{M}_\odot$ and $5.6\,\mathfrak{M}_\odot$, respectively. From spectrophotometric measures, Lee and Wright determine Δm as 1.9 mag.; from the light curve, Popper computes it to be 2.2 mag. In their graded catalogue, Koch et al. give values of a_g and a_s (radii of larger and smaller components) as 0.166 λ and 0.0024 λ, respectively, but with a rating corresponding to the lowest possible reliability. The point out the need for further observations both during and outside eclipses and mention the wavelength dependence of the phases of internal and external tangencies. They conclude that, "The system is known far more poorly than is generally believed".

31 Cygni. Although this star has been given an official variable star appelation of V 695 Cygni (as well, of course as BD and HD numbers and other designations) it is almost universally referred to in the literature by its Flamsteed number. The system is one component of a visual binary. It also is covered in the excellent earlier summaries by Wilson and by Wright. In many ways, it is remarkably similar to ζ Aurigae. The spectrum of the cooler star in each system is classified as K4Ib; however, the radius of that in 31 Cygni is one and one-half times as large and in absolute magnitude it is nearly one magnitude brighter the cooler component of ζ Aurigae. The blue component of 31 Cygni is somewhat hotter than that in ζ Aurigae. Wright, after some discussion, lists B4V as compared to B6+V for ζ Aurigae with a temperature of 17,600°K. The components are more widely separated with a period of 3784 days contrasted to 916 days. The diameters found from the spectroscopic data do not agree closely with those found from the eclipse data, but Koch et al. (1970) consider the latter to be extremely poorly determined for each system.

This system was discovered to be a spectroscopic binary in 1901 (Campbell, 1901). The spectrographic orbit was first determined by Vinter-Hansen (1944), although Christie (1936) has previously obtain provisional spectrographic elements. McLaughlin (1950a) announced atmospheric (ζ Aurigae-like) effects had been observed in 1941. The 1951 eclipse was extensively covered. Discussions have been given b

McLaughlin (1952a, 1952b), McKellar et al. (1952), McKeller and Petrie (1958), Larsson-Leander (1953, 1957), Wright and Lee (1956), Ezer (1961) and others. In fact, each of the first four numbers of Volume XI of the Victoria Publications discusses some aspect of the spectrum; together they virtually provide a monograph on the spectroscopic features. The radial velocities during the atmospheric phases of the eclipse behave in a complicated manner. The interpretation strongly supports the idea of the chromosphere of the K star consisting to a great extent of "discrete moving masses of gas". The work by Larsson-Leander and by Wright and Lee were extensive spectrophotometric studies; among other findings, the light-ratio of the components was determined at various wavelengths.

The concept of an atmospheric eclipse seems well established. Wood (1953) used photoelectric observations by himself and by Carpenter to show that, as in ζ Aurigae, the fractional light loss at any phase, depends on the wavelength of observation. Underhill (1954) studied the development of the Fe I lines during ingress. She derived the light curve of the B-star from intensity tracings and found that it was reduced for several days prior to the first contact.

At the 1961 and 1971 eclipses, extensive photoelectric and spectroscopic observations confirmed fully the atmospheric nature of the eclipse. The photometric observations made during the 1961-62 collaborative campaign have been compiled and discussed by Gyldenkerne and Johansen (1970) and limiting values of the radii have been computed by Johansen (1969). No really major changes in the interpretation of the system were found; the atmospheric nature of the eclipse was strongly confirmed and the advantage of the use of a common narrow band filter system was clearly demonstrated. Saito and Kawabata (1976) have interpreted their observations of the Ca I 6572 Å line in terms of an expanding circumstellar gas cloud.

Florkowski (1975) has observed the system on two occasions at 3.7 cm and 11 cm and found no detectable radio emission. Similar negative results were found for ζ Aurigae and VV Cephei. Observations at the Villanova Observatory (Bull. Amer. Ast. Soc. 7, 215, 1975) reported a gradual increase of light (about 5%) from the end of primary until a year later. Continued observations of this and similar systems in the intervals between eclipses probably would aid greatly in their interpretation.

32 Cygni. This system differs in a number of ways from the preceding ones. For one thing, the orientation of its orbital plane relative to the solar system is different, so that the eclipses are not of the nearly central type with long intervals of totality. Different authors have estimated values of the inclination ranging from $72°$ to $82°$. Since the computed absolute values of the masses depend on the $\sin^3 i$, this means they are not well determined. The extent of the atmosphere and the duration of totality almost certainly varies from eclipse to eclipse, as well as with wavelength of observation. This, combined with the uncertainty in the inclination makes it exceedingly difficult to determine reliable values for the radii. On the other hand, this geometry means that the "atmospheric" effects connected with the chromosphere are of greater relative duration and can be more thoroughly studied. Wright lists the spectra as B4 IV-V and K5 Iab in his review article and states that the latter is not much different from the other K-stars in this group.

A radial velocity curve was observed and spectroscopic elements computed as early as 1918. The similarity to ζ Aurigae was recognized from spectra taken in 1949; the nearly grazing totality was also recognized then (e.g. McLaughlin, 1950b). Absolute dimensions were computed by Wellmann (1953) and by Wright (1952) from observations in the 1952-53 eclipse. They agreed in finding masses of about 21 $\mathfrak{M}_\odot$ and 8$\mathfrak{M}_\odot$ for the K and B stars, respectively, although, for reasons given earlier, these values cannot be considered as highly reliable. At this eclipse, also, both Herczeg (1956), and Wood and Lewis (1954) found evidence of intrinsic variability not directly connected with eclipse effects; this was confirmed by observations by Botsula (1962) from observations running from 1957 to 1961. A good deal of attention was given to determining the exact duration of the partial phases. This naturally proved difficult since, as we now know, this varies from eclipse to eclipse and depends on the wavelength of observations as well.

Later eclipses had many observers both photometrically and spectrographically, who established the generalities previously summarized. We should mention in particular the extremely careful study of the Ca II-K line made by Wright (1969) at the 1965 eclipse. Many components of this line were observed on spectra taken within two months of totality; some were as much as 1.5 Å from the principal line. In extreme cases, their lifetimes ranged up to several weeks. At this eclipse also, the light-ratio of the continuum at λ 3900 Å of the B to the K star was determined as 0.4; this is considerably less than that for either ζ Aurigae or 31 Cygni.

Johansen et al. (1970) made a thorough analysis of UBV observations in the 1959, 1962, and 1965 eclipses and found the atmospheric nature clearly demonstrated. The 1968 eclipse was somewhat less intensively observed although it was not neglected (e.g. Sato and Saito, 1969), while that in 1971 was the object of an internationally coordinated program (Wright, 1972). There included both wide band (Saito et al. 1972) and narrow band (Magalashvili and Kumsishvili, 1974; Bloomer and Wood, 1974; Sato and Saito, 1973) measures. The 1974 eclipse again was somewhat less thoroughly observed although narrow band photoelectric observations were made.

Two other interesting series of observations should be mentioned. Guinan and McCook (1974) have published a series of photoelectric observations at Hα and Hβ made outside eclipse and reported on variations in these regions. Doherty et al. (1974) have reported on ultraviolet observation made in eclipse at seven wavelengths from the OAO-2 satellite. At several of these the K-star was brighter than expected and this raised the question of emission in an extended envelope.

Galatola (1972) has presented an interesting model of the system based on the blue and ultraviolet observations of Herczeg and Schmidt (1963) and of Johansen et al. (1970) during the 1962 eclipse. The model has an opaque core surrounded by a semi-transparent envelope. He found an inclination of 82° and, at this eclipse, a value of 1.87 $R_\odot$ for the radius of the B-star. The K-star's radius was found to be 120 $R_\odot$ for the blue light and 123 $R_\odot$ in the ultraviolet. It will be of interest to see if this model can be applied in detail to other systems of this type.

VV Cephei.   This system shares many, but not all, of the common characteristics of the three just discussed. While the cooler component

of each of the preceeding systems was clearly a K-type, that of VV
Cephei has been classified as M2 Ia-Iab. According to Wright's summary, the hotter component is classified Be; although Wright and
Larssen (1969) state that the only real evidence is a strong ultraviolet continuum. Even allowing for the many uncertainties involved
in the determination of the radii, that of the supergiant component
of VV Cephei must be at least five, and possibly ten or more, times
larger than that of its counterpart in any of the other systems. A
ten-fold larger radius of course means a volume 1000 times greater.
High dispersion spectrographic studies indicate steady mass loss from
the primary. Other systems which appear similar from spectrographic
studies have periods of the order of ten years or longer and masses
of the cooler components of more than thirty times that of the Sun
(Cowley, 1969).

Although the system was known to be unusual as early as 1907, when
Miss Cannon found an M2 spectrum with hydrogen emission, the suspicion that it was an eclipsing system similar to ζ Aurigae was first
expressed by McLaughlin (1936) - an astronomer who contributed very
largely to the earlier development of the studies of such stars. The
first photometric study was made by Gaposchkin (1937a, 1937b) who
used photographic estimates from plates taken in three minima and derived light elements; he also found intrinsic variations in the M-
component. Gaposchkin also called attention to the extremely small
value of k, the ratio of the radii. A great deal of spectroscopic
work was carried out in the late 1930's, almost all of it following
McLaughlin's announcement.

An extremely complete summary has been given by Goedicke (1939a,
1939b). He confirmed the extremely large mass-function earlier found
by Christie (1933). From time to time, other authors have summarized
the work. For example, McLaughlin (1952) described the spectroscopic
effects of the atmospheric eclipse and contrasted it with similar
systems. Shortly thereafter (McLaughlin, 1954), the K-absorption line
intensified and developed remarkable complexity and variation; two
lines of Ti II in absoprtion became conspicuous. Wright and McKellar
(1956) and McKellar et al. (1957) presented a detailed description of
the chromospheric spectrum during the ingress and totality of the
eclipse of 1956-57. Keenan and Wright (1957) discussed carefully the
matter of spectral classification of the cooler component and finally
gave: M2+ Ia-Iab.

A number of observers made photoelectric observations during and
near this eclipse; the most extensive set was by Larsson-Leander
(1957, 1959, 1961). He reported that the light and color curves show
variations in light of the M component in cycles of about 100 days
with amplitudes of about 0.15 mag.

This system has also been studied astrometrically, but the results
have added to rather than solved some of the puzzles. Discrepancies
which exist between the spectrographic data and that deduced from
photometric and astrometric data have been discussed by Fredrick
(1960). One feature of the photometry during eclipse was the appearance in his observations of pulse-like phenomena on four occasions:
the amplitude increase with decreasing wavelength, but Deutsch found
no traces of the B-star on a spectrogram taken almost simultaneously
with one of them while three occurred before the time of third contact as determined by Wright. Fredrick suggest that the source is
in a circumstellar shell or a florescence in the atmosphere of the

M-star caused by a brief increase in radiation from the B-star; to date, other observers have not found such pulses.

Another peculiarity is that the absolute visual magnitude of the M component as determined from the derived parallax of +0.005 and the apparent magnitude during totality (allowing for absorption) is -1.36 mag. Even a parallax of 0.004 would give only an absolute magnitude of -1.84. This is completely discordant with the absolute magnitude of a M2+ Ia-Iab star. Fredrick tries to argue that the magnetic field discussed by Babcock (1953, 1958) can sharpen the absorption lines. This, however, does not explain why the many series of spectrographic observations have reported no changes coinciding with changes in the magnetic field. The spectrographic material is so extensive and it is the work of so many experienced observers working at various dispersions that strong evidence is needed to believe it is so systematically wrong. The absolute parallax is small with a quoted probable error of nearly half its value (π_a = +0.0054 ± 0.0023) and is derived from an even smaller relative parallax (+0.0042) itself found by corrections to the least squares value of the x coordinate. Assumption of no systematic error is, of course, inherent in all least squares treatment. If the astrometric parallax made the system systematically much too bright, thus implying that it (the parallax) was much too small and should be increased, we might feel concern, but the reverse is true. Serious uncertainties in the dynamical parallax, such as that arising from uncertainties in the wavelength response to a refractor in determining the photo-center, especially in view of the large changes of light and color, make us reluctant to accept a result of so small an angle as 0.0026 as a lower limit. Until an independent astrometric determination is made, we prefer to rely on the spectroscopic data for the values of the absolute magnitudes. Indeed, an absolute magnitude of -5.5 is not excluded by the astrometric data. The corresponding distance would give a parallax π = .001, which is only 1.3 standard deviations from the observed value π = .005.

Solution of the light curves in the case of extended atmospheres. From the preceding discussion it is obvious that the application of methods used to solve light curves of more conventional systems will give at the best only crude approximations in the systems we are discussing. The "radius" of the supergiant stars, for example, depends on the wavelength in which the observations are made. Clearly, different techniques must be used if we are to determine with maximum precision the parameters involved and, almost of equal imporatance, get a reliable estimate of the precision with which they are determined. Although the term "extended atmospheres" normally used more extensively than in this section has been, we will confine the discussion primarily to stars of the ζ Aurigae or VV Cephei type. A good description of the work to date in the more extensive field including a listing of earlier papers has been given by Cherepaschuk (1974).

A study of systems with extended atmospheres, chiefly with the aim of developing equations to permit calculation of fractional light loss, was started by Linnell (1958). In his paper, Linnell summarized some of the earlier work discussing gaseous streams or extended atmospheres (e.g. Struve, 1949; Deutsch, 1956; Ovenden, 1956; and others) and the limited amount of theoretical treatment of light curves which had started with the treatment of ζ Aurigae by Mènzel (1936).

The existence of an extended atmosphere means, of course, that the number of elements needed to describe the system be increased at least by one and by more than this if one parameter is not sufficient to describe the effects of the atmosphere on the observed light from the eclipsed star. Some assumption as to the nature of the extended atmosphere is needed, and it is by no means certain that different assumptions may not give predictions which cannot be distinguished by the observational evidence. This is particularly true since interpretation of some of the spectrographic material strongly suggest that the atmospheres of the giant stars are not in hydrostatic equilibrium.

The treatment first considered monochromatic observations and a star with the conventional cosine law of limb darkening. Ellipticity and reflection effects were ignored as well as the possibility of stimulated emission in the extended atmosphere. Finally, an extinction law, $dI/I = \kappa\rho\,ds$ was used; where κ and ρ are the mass absorption coefficient and the density respectively. In a later work, Linnell (1961) published tables giving the fraction of the light loss of an undarkened spherical star undergoing occultation by a larger star with an extensive atmosphere. Another early effort for solving light curves in the case of an atmospheric eclipse was made by Lukatskaye and Rubashevski (1961).

Various treatments have been given of light curves of "atmospheric eclipses" using the concept in a much broader sense than we are considering here and including Wolf-Rayet stars, W Ursae Majoris systems, and others. As one example, there are the computer programs developed by Cherepaskchuk (1973). This is a direct method which uses FORTRAN to give both the elements and their respective errors.

In conclusion, we note that while the general nature of these systems now seems well established, much further observational work is needed. Simultaneous spectrographic, photometric, and spectrophotometric observations are required. The desirability of the use of identical sets of intermediate or narrow band filters for the photometric work has been clearly demonstrated. Observations should be made in the intervals between eclipses, as well as near and during the, and the work of different observers should be coordinated to the maximum possible extent.

This has been a brief summary of the tremendous amount of work carried out on these systems in recent decades. Lessons have been learned which almost certainly will be applied to future studies. The value of the use of relatively narrow band color filters, as nearly identical as possible for different observers, has been thoroughly justified by the results and no doubt will be applied extensively in the future. The turbulent conditions existing in the extensive atmospheres of the supergiant components, especially that portion exposed to the radiation of the B-component, present situations of extreme complexity. Our chief and almost only chance for gaining better understanding lies in taking full advantage of those fortunate times of eclipse, when we can observe the changes which occur when the radiation received from the hotter star passes through successively greater layers of the atmosphere of the cooler star. If past performances are signs of the future, full advantage will be taken of these opportunities by many interested astronomers.

CHAPTER 10

INDIVIDUAL BINARIES OF PARTICULAR INTEREST

1. ALGOL (β PERSEI)

1.1.   One might think that a system which has been as intensively
studied as Algol would present no problems of interpretation. Yet
such is far from the case. Approximately 1000 papers, some of them
rather trivial but many substantive, have been published concerning
it. Yet, in the introduction to a paper published in 1971, the authors (Hill et al., 1971) write, "remarkably little is known of the
individual components of the system". It is more than a double star.
Different authors have at times suggested as many as 6 components in
the system, but at the time of this writing, it seems that three are
all that are necessary to explain the observed phenomena.

It is the earliest known eclipsing system, both in original discovery and for which the systematic studies have been made. Historically, it had been a "first" in various studies. The precise timing
of its minima by Hall (1939) in two colors was probably the first
strong evidence against the Tikhoff-Nordmann effect. This report
that minima observed in red light preceded those in blue indicating
different velocities of light in different wavelengths had been "established" in 1909 and "confirmed" as late as 1935. In a spectrophotometric study, Hall timed the minima as observed at λ 5500 Å and compared these times to those observed at λ 8660 Å. They occurred simultaneously with an uncertainty of not more than 3 minutes; other work
on much weaker evidence has reported discrepancies up to 13 minutes.

Algol was the first eclipsing system to have a light curve observed
in the infrared and the orange spectral regions (Chen and Reuning,
1966). It has been observed from artificial satellites in the ultraviolet. Flares have been observed in radio emission and it is also
a source of intermittent X-radiation. A detailed discussion of its
observed period changes and their proposed explanations would itself
merit a chapter or two. This is the more remarkable because it is
apparently a normal member of the "Algol-type" systems, but one which
happens to be close enough to us to be studied in some detail.

The story of the discovery and interpretation of its variability
by Goodricke in 1783 has been mentioned in Chapter 1. Yet, Goodricke
was by no means the first to observe the light variation. More than
a century earlier in 1669 Geminiano Montanari had recorded the fact

that the light varied at times.   Even he was far from the first.
Roughly 2000 years earlier, the Chinese had related the observed
brightness changes of Tsi-Chi (積屍) or Algol with death of the
population or good or bad fortune; this was recorded in Hsing Ching
(星經) written in the period from 205 B.C. to 25 A.D.

   The work on Algol since Goodricke's time is far too extensive to
permit any but the most sketchy summary of the chief developments.
In 1889 Vogel and Scheiner made spectroscopic observations which
established the hypothesis of its binary nature.   In 1910, in the
earliest days of astronomical photoelectric photometry, Stebbins
(1910) used a selenium conductive cell to detect the shallow secondary minimum, and to detect variations in the light between eclipses.
The first light curves in two colors were reported by Maggini (1918),
who used a wedge photometer and filters to observe at λ 6450 Å and
λ 4120 Å.   The first complete photoelectric light curve was published
in 1921 (Stebbins, 1921).   In each of his papers, Stebbins also published photometric solutions and computed absolute dimension.   Examination of 336 spectrograms suggested long period variations in the
radial velocity of the primary component.   As a result he examined
all the observed light minima from 1852 to 1887 and found a fluctuation of the period.   Shapley (1915) made a solution using the observations by Stebbins.   Hellerich (1922) presented spectroscopic elements and made a careful comparison with earlier work; he also compared
the spectroscopic and photometric data in general.   The first suggestion of a third body in the system seems to be due to Stebbins (1922)
as a result of a study of the period variation.   Another careful
spectrographic study was made by McLaughlin (1924) who computed absolute dimensions and also confirmed the variations of velocity of the
center of mass of the eclipsing pair as reported by Curtiss (1908).
We cannot in any reasonable space record in detail the tremendous
amount of work done on the system during the following years.

1.2.   Algol A is an apparently normal B8 V star, although as noted
earlier emission features at certain times have been reported.   Algol
B has not been detected spectrographically.   From the solutions of
the various light curves it seemed to be a subgiant of late G or
early K class. Algol C, observable only near the middle of primary,
had been classified as an Am star (Fletcher, 1964).   The eclipsing
period is approximately 2.87 days; the period of AB and C around their
common center of mass is approximately 1.86 years.   Algol C has not
been (and is not likely to be) detected as a visual companion in the
classical sense but it has been observed as an astrometric binary,
and Labeyrie et al. (1974) believe that they have resolved it using
speckle interferometry.

   Interest developed in the long suspected third star in the system.
As early as 1923, narrow lines, later attributed to the third component, were observed on spectra taken during primary.   Struve and
Sahade (1957b) studied high dispersion spectrograms and found many
narrow lines attributable to the third component.   They classified
this as late A of early F with a luminosity class of IV or V, and
found a magnitude difference between components A and C of approximately two magnitudes.

   Ebbighausen and Gange (1963) discussed the motion of the eclipsing
pair in the triple system.   Hill et al. (1971) have analyzed an extensive amount of spectrographic observations taken between 1937 and
1941.   They derive masses of $3.7\mathfrak{M}_\odot$, $0.8\mathfrak{M}_\odot$, and $1.7\mathfrak{M}_\odot$ for components
A, B and C respectively.   A somewhat speculative section discusses

the evolutionary state of the system in terms of mass exchange.

Bachmann and Hershey (1975) have discussed the orbit of Algol AB, C combining astrometric, photometric, and radial velocity data. This combined treatment gives masses of 5.3 $\mathfrak{M}_\odot$ for Algol AB and 1.8 $\mathfrak{M}_\odot$ for Algol C. The value for AB differs appreciably from the value of 4.5 $\mathfrak{M}_\odot$ found by Hill et al. (1971) from spectrographic observations; the spectrographically determined value of 1.7 $\mathfrak{M}_\odot$ for Algol C agrees well with the above.

The idea that the Algol system might contain more than three bodies has rested largely on interpretations of supposed periodicities in the period variation. Many extensive studies were made of the period changes beginning with Chandler (1901). References to these have been given by Frieboes-Conde et al. (1970). There have been various inter-pretations. One of the most common has been to interpret apparent periodicities in the (O-C) diagrams as light time effects caused by other members of the system. This led originally to the suggestion of a third member of the system, and then of a fourth, a fifth, and on some occasions a sixth member. In some ways, the situation was similar to the creating of more and more epicycles in order to pre-serve the Ptolemiac theory of planetary motions. However, as obser-vations continued some of the periodicities failed to repeat, and doubts as to the reality of components D, E, and F became increasingly strong. Finally, in 1970 Frieboes-Conde et al. combined a period study with astrometric observations to destroy three "stars" in a single paper and to reduce Algol to a triple system. We shall summar-ize their rather extensive treatment.

At the time of their study, the periodic terms in the (O-C) dia-grams had been reduced to three. One had long been called the "great inequality". Its amplitude was more than three hours and its "period" about 180 years. Two open questions were: (1) whether a period of this length can be considered well established even in a system which has been observed as long and as faithfully as Algol, and (2) whether it was properly interpreted as a light-time effect caused by the motion of the center at mass of Algol ABC around an unidentified fourth body.

Another periodicity suggested from the light-time was that of 32 years which had been attributed to Algol E. An exhaustive discus-sion of the astrometric data fails to support the alleged 32 period as caused by orbital motion. Further, the required mass for Algol E suggests a brightness comparable to Algol C and raises the question of why the spectrum has never been observed during eclipse. The authors conclude that the postulate of a fifth component is "very doubtful" and prefer the interpretation of the changes as caused by apsidal motion within the eclipsing system. This is rather strongly confirmed by Hill et al. (1971) who plotted against epoch the spectro-graphically determined values of ω and found a slope corresponding to a period of 32 years.

In considering the "great inequality", Frieboes-Conde et al. ex-panded the work of Ferrari (1934) who had examined all the observa-tions used in deriving the FK3 position of the star, and who had found no indication of orbital motion with a 180 year period. In a careful discussion, the authors conclude that Algol AB does not show the orbital motion required. They find its motion since about 1875 to be perfectly rectilinear. They conclude that observations before that date -- the meridian observations go back to 1754 -- are not sufficiently accurate to be included.

Nevertheless, the light-time effect remains and requires explanation. To invoke apsidal motion in this case would require an orbital eccentricity of at least 0.15 and this is forbidden by both the photometric and the spectrographic observations. This raises the question as to the validity of the periodic nature of the long range variations. The range of precise observations is only a fraction of this interval and residuals between observed and computed times are already large. After allowing for the known changes discussed above, it is found that the eclipsing period occasionally undergoes sudden changes -- a common feature in similar systems -- but remains constant in the intervals between. Consideration of the epochs determined from photoelectric observations suggest a ΔP of + 3.5 sec about 1944 and one of -2.1 sec in or about 1952. Thus, three components are all that are needed to explain the observed period changes to date.

1.3. The light curve of Algol presents certain difficulties in the analysis. Some factors contributing to this are the shallow depth of secondary minimum at all but the longer wavelengths, the partial eclipses, and the light from a third component which is more luminous and of higher temperature than Algol B. We mention a few of the developments.

An extensive study by Herczeg (1960) not only presented what was probably the most reliable photometric solution to that time, but also suggested systematic changes in the depths of primary eclipse.

In 1966, Chen and Reuning (1966) published light curves made with a lead sulphide detector and centered at λ 0.60 μ and λ 1.6 μ. The elements they derived lie within the range of those given by earlier observers in the shorter wavelength regions. They found that the pronounced secondary minimum, strikingly evident in the 1.6 μ observations, greatly enhanced the precision with which the orbital elements could be determined, and suggested that observations of similar systems in the infrared would be of considerable value.

Cristaldi et al. (1966) used an interference filter of half width 65 Å centered on the H_α line. Their derived elements agreed well with those found earlier but the search for evidence on istability manifested by change in the equivalent width of this line was inconclusive. The authors emphasize the importance of a systematic observing program isolating spectral features such as the H_α and H_β lines in which changes would be indicative of stellar instability The "b_1" solution given by them was selected by Koch et al. (1970) in their graded catalogue as the most suitable to that date. Although the fit in secondary eclipse is not good, the analysis has removed from all the observations the light of the third component. It was given grade "C", which is fairly high in view of the rather severe criteria applied by Koch et al. in their grading system.

We have already referred to Hall's 1939 publication of results of photoelectric observations in the visual and near infrared. Six color photoelectric observations were made by Stebbins during 1949-51, but have only recently been published (Stebbins and Gordon, 1975).

Wilson et al. (1972) presented two narrow-band and one broad-band photoelectric light curves and used the recently developed Wilson and Devinney method for an analysis. Grygar and Horak (1974) solved six light curves ranging from λ 3700 Å to λ 16,000 Å using the sphere-ellipsoid and the sphere-sphere models. Both this paper and the one by Wilson et al. summarize the more recent of the preceeding photometric studies. Other workers were far from idle.

Struve (1953) confirmed Morgan's observation that Mg II λ 4481 Å became double near mid-eclipse. Tests for polarization were negative. Various authors used various techniques to determine the limb darkening of the hotter component (e.g., Cayrel-de Stroble, et al. 1955). Cayrel-de Stroble (1955) measured the depths of the Balmer lines and the Ca II - K line at various phases, found that the depths increased during eclipse, with the increase beginning a little before the start of the geometric eclipse and continuing a little after its end; she suggested an extended atmosphere.

As early as 1923, narrow lines, later attributed to the third coponent, were observed on spectra taken during primary. Struve and Sahade (1957a, 1957b) studied high dispersion spectrograms made near minimum light and near phase 0.75 days. They discussed the emission features found at the latter, and interpreted them as effects of a gas stream. The double lines of Mg II λ 4481 Å found near center of primary and first reported by Morgan in 1935, proved more difficult to explain.

Sahade and Wallerstein (1958) studied the spectrum in the near infrared and found no evidence for the spectrum of Algol B; they suggest that earlier reports of such a spectrum are probably accounted for by the blending of the lines of the A and C components.

1.4. We have already mentioned the emission at a quarter point of the light curve observed by Struve and Sahade. Other evidence exists which indicates that at least at times circumstellar material exists in the Algol system. From measures of the equivalent width of the HeI λ 4471 Å line as a function of phase, Fletcher (1964) deduced the presence of circumstellar helium which was not distributed symmetrically around the line of centers.

Additional evidence for the existence of circumstellar material and some idea of its extent was found by Guinan et al. (1976) from light curves centered at H_α and H_β and using both wide-band and narrow-band photometry at each wavelength. Noticeable differences exist between the narrow-band and wide-band curves, especially in H_α near the shoulders of primary. The analysis of the nature of these differences suggests that the material extends to a distance of approximately $2.6 \mathfrak{M}_\odot$ times the radius of the hotter component and quite possibly fills completely its Roche lobe.

We might expect some variation in infrared excess with phase, but the published evidence at the moment of this writing is contradictory. However, studies of the radio emission are in better agreement. On 25 February 1972, IAU Circular No. 2388 carried a report of radio flare activity by Hjellming, Wade and Webster and also a report by C. T. Bolton on optical spectrum variations which possibly might be correlated with the radio emission. Algol had earlier been detected as a quiescent radio source by Wade and Hjellming. Further observations in the radio range and the nature of Algol as a radio source have been discussed in Chapter 5.

It might be noted that a serious error has crept into the literature in the discussion of Algol as a radio source. Frieboes-Conde et al. are frequently cited as giving the value of 0.2 A.U. for the semimajor axis of the eclipsing system - an obvious impossibility if Kepler's third law can be trusted. What Frieboes-Conde et al. did was to cite Eggen and Pavel as giving this value for the distance from the center of mass of the Algol ABC system to the center of mass of a system containing the hypothetical fourth star Algol D, now believed to be non-existent.

We might also note two final items. Borra and Landtreet (1973) attempted to observe a magnetic field but their single observation did not detect one. Pfeiffer and Koch (1977) included Algol in their extensive study of linear polarization of close binaries and found no detectable polarization thus agreeing with the earlier results. This latter finding, combined with the results of Guinan, the radio emission, etc. indicates that the presence of circumstellar material will not necessarily be indicated by partial polarization of the emitted radiation.

1.5. The X-ray observations have already been described in Chapter 8. There have also been published a few studies based on satellite observations in the ultraviolet. Eaton (1975) published photometric observations and spectral scans taken from the Orbiting Astronomical Observatory (OAO-2) in August 1969 and in July 1972. Light curves were obtained in wavelength regions centered at λ 2930 Å, λ 1920 Å and λ 1550 Å. This was a pioneer attempt and except for the observations at λ 1920 Å, the photometry was not of high quality, as the author himself states. The light curve was incomplete with no coverage between phases 0.2 and 0.5. An analysis was made to establish effective temperatures and luminosities of the three components although the values obtained for Algol B and Algol C were "rather uncertain". Nevertheless, this gave an insight into what might be accomplished by such observations in the far UV, especially in a study of the wavelength dependent parameters of the system.

Chen and Wood (1975, 1976) obtained observations from the Princeton Experiment Package aboard the Copernicus satellite on September 7, 8, and 9, 1973 and January 4, 1974 and have discussed these so far in two papers to date. One was a study of the Lyman-α line within and outside of primary eclipse. This showed no emission features and is consistent with what would be expected for a normal B8 star. The second paper discussed variation in the Mg II resonance lines near λ 2800 Å. Near the center of primary minimum these appeared to be doubled -- a feature discovered much earlier in Mg II at λ 4481 Å. A possible explanation is the effect of emission from Algol B which becomes detectable only when almost all of Algol A is in eclipse. Kondo et al. (1972) had earlier found emission features in this doublet for stars of spectral types F-M. Use of the relation between emission line width and absolute magnitude relation given by Kondo et al. (1975) gave an approximate value of 4.3 for the absolute magnitude of Algol B which agrees with the published ones as computed from other data. However, Struve and Sahade had explained the doubling in Mg II λ 4481 Å by the blending of the lines of components A and C, so the peculiar behavior at these lines near mid-eclipse needs further study. These same Copernicus observations were used by Chen et al. (1977) to analyze a light curve at λ 3428 Å. The results indicate a temperature of Algol B higher than 5000 °K.

Thus at the time of these observations, Algol appeared in a quiescent state. The only evidence of unusual activity were the relatively small irregular changes in profile of certain lines during eclipse which could be interpreted as modification by circumstellar gas. Yet other evidence, such as the period changes and radio flaring, suggests that this is not always the case. Indeed, later observations of Mg II

by Kondo et al. (1977) made using a balloon-borne spectrometer have
given evidence of mass flow and other peculiarities. These observations were made on October 9-10, 1974. It will be of interest to see
whether later observations show period change near this epoch.

Thus Algol, the first discovered and one of the most frequently
observed eclipsing binaries, still is not completely understood and
merits further observation. We might add that the above discussion,
while intending to give a general picture of the various features of
the system, is by no means complete in the sense of mentioning every
useful contribution. When such have been well summarized in later
publications mentioned here, reference has frequently been omitted.

In general, we may say that our present picture is that of a system still undergoing slow mass exchange and probably associated mass
loss. The exchange or loss is not of the steady, systematic sort
shown by computer stars, but seems to occur during certain intervals --
something like U Cephei events, but on a much smaller scale. It is
interesting to speculate on whether there is a rough periodicity to
these something like the sun-spot cycle and whether they are indeed
associated with the existence of stellar spots, flares, and related
activity. It will be of greater interest to see what solid information and surprises future observations will bring.

2.   β LYRAE

The 13-day period binary β Lyrae (Sheliak to the Arabs), as mentioned already in Chapter 1, was discovered as a variable star in
1784 by the deaf mute amateur John Goodricke (1785). It is the object
on which there are more papers written than on any other individual
star.

William Herschel had suspected a few years before its discovery
as such that either β or γ Lyrae were variable, and the first light
curve was derived by Argelander in 1859 from eye estimates. There
are spectra taken as far back as 1886 at the Harvard Observatory and
the first spectrographic orbit that was published was due to Belopolsky (1893), as mentioned in Chapter 1.

We should make special mention of the model that Myers (1898) ar-
rived at from an analysis of Argelander's light curve and Belopolsky's
velocity curve. Myers concluded that the system was formed by two
similar ellipsoids of revolution with masses of 21 and 9.5 solar
masses, respectively, moving in nearly circular orbits. He further
stated that the mean density of the system is comparable with atmospheric density or the system is "in a nebulous condition". Moreover,
that there should be a "powerfully absorbing envelope about the nuclei
of the masses" which "would arrange itself about the combined mass of
the system so that portions of equal density would be in equipotential
surfaces".

β Lyrae has puzzled - and is still puzzling - astronomers since its
spectrum was observed for the first time. It was the pet star of
Otto Struve, who devoted practically all his life to try to explain
its spectrum and to understand the kind of object we are dealing with.
In one of his many review papers on the star, Struve (1958) remembered
that he "watched its changes when on night duty during the First World
War and.......have been puzzled over its peculiar spectroscopic behavior

at frequent intervals during the past 35 years".

A great advance in our understanding of β Lyrae came about when
Struve (1941) published his thorough study of the star in volume 93
of The Astrophysical Journal and Kuiper (1941) tried to establish the
theoretical basis for such an interpretation. The interpretation proposed by Struve was the starting point for a great revolution in the
field of close binaries. Unfortunately, at the time it was believed
that every star strictly adheres to the mass-luminosity relation and,
as a consequence, the derivation of the masses was based on an assumption that we now know is not always true. This delayed further advances for some fifteen years or so.

If we have at hand the excellent reproductions in Trans. American
Phil. Soc. 49, 1 (1959) we could see that the spectrum of β Lyrae is
characterized by the presence of both absorption and emission features.
The absorption lines can be separated into three different sets, namely,

a)   a group of lines, let us call it set a, which corresponds to
the normal spectrum of a B8 II star;

b)   a second group of lines -set b - which shows the effect of diluted radiation, the strongest ones in the photographic and near infrared regions of the spectrum being the triplet lines of HeI at 3888
and 10830 Å;

c)   a group of extra lines -set c - which are present only during
primary eclipse and are usually called "satellite lines" after Baxandall
(1930) who discovered them; before mid-eclipse the satellite lines appear at the red side of a number of the stellar absorption lines, and
after mid-of principal eclipse, they appear displaced shortwards. They
are referred to as "red satellite" and "violet satellite" lines, respectively.

In the emission spectrum we distinguish the set of broad lines that
are seen in the photographic and visual region of the spectrum, and
the P Cygni profiles that are observed in the ultraviolet.

The B8 II spectrum corresponds to one of the components of the system and yields a radial velocity curve from which the orbital elements
can be derived. If one considers the spectral lines that are free
from blends, the orbital elements correspond to a semiamplitude of
185 km sec^{-1}, a sin i = 32.9 x 10^6 km and f($\mathfrak{M}$) = 8.5 $\mathfrak{M}_\odot$, the computed
eccentricity being e = 0.017 (Sahade et al, 1959). They have been
derived from the lines of SiII.

The velocity curve from H, and perhaps the one from HeI, at about
secondary minimum, departs slightly from that of Si II, in the sense
that the velocities of H and HeI fall below those from Si II by some
6 km sec^{-1}. Another departure from the normal velocity curve appears
at primary eclipse due to the rotation of the B8 component; which is
of the order of 45 km sec^{-1}, as measured from the width of the H absorption lines. Rossiter (1933) was first in detecting the rotational
disturbance on the velocity curve of β Lyrae. After mid-eclipse an
additional effect appears superimposed upon the effect of the rotation
of the B8 star. This additional effect, which we have commented upon
in Chapter 5, is variable and the observations made have detected a
range of some 10 km sec^{-1}; they have been interpreted as caused by
electron scattering (Struve, 1958). Struve (1941) has also attributed
to electron scattering the fact that during primary eclipse, more pro-
nouncedly after the phase of minimum light, the lines of the B8 star
become broad and hazy.

The light curve of β Lyrae shows a continuous variation of brightness with phase and two minima which, in the photographic region, are of about 0.9 mag. and 0.4 mag. deep, respectively. In such a case, one would argue that the difference in the depths indicates different surface brightness in the two stars. If the one which is eclipsed at primary minimum is, as the velocity curves suggest, the B8 II (earlier classified as B9) star, then the companion should be a later type object, actually an F star. Spectrographic observations in the near infrared have failed, however, to detect the presence of a late type companion (Hiltner, 1947; Struve and Sahade, 1958).

It is possible to solve the light curve of β Lyrae by assuming that the "cooler" star is either larger or smaller than the B8 companion. The earlier observers adopted in their solutions the first possibility but this was at variance with what the spectrographic observations indicate. Actually, the principal eclipse is almost central (i $\sim$ 80°) and yet the spectrum of the B8 component is seen at all phases even at minimum phase when the light has dropped by nearly one magnitude. Therefore, the eclipsing star at principal eclipse - the companion to the B8 star - must be the smaller of the two. Such a situation is the one which has been considered for the solutions of the light curve since Gaposchkin (1938) worked out his.

The light curve shows changes during eclipse particularly after mid-eclipse. Larsson-Leander's (1969) photometric results of the 1959 international campaign organized by Struve to observe β Lyrae, show these changes extremely well; they could amount to tenths of a magnitude or so. Moreover, the observations of times of minima of β Lyrae have disclosed the fact that the period of the light changes increases with time at the rate of $\sim$ 9^s.5 per year. Argelander already realized this and gave the times of minima of the star with a third power term.

The change in period in β Lyrae has been interpreted in terms of mass loss and the derived rate is $\sim$ 10$^{-5}\mathfrak{M}_\odot$ yr^{-1}, which led to the conclusion that the star is losing mass at a rapid rate.

As we have already mentioned in Chapter 5 (3.2.1.), one of the big puzzles that were posed by β Lyrae was the question of how we could have the light elements that characterize the system, when a relatively small star is in front at primary minimum. This question was answered by Huang (1963) who suggested that the companion to the B8 star is surrounded by an opaque, flat disk. The existence of such a flat envelope is also suggested by the polarization measurements. We do need to repeat what we have already discussed in Chapter 5 (3.3).

β Lyrae was the first close binary with a peculiar spectrum that was investigated. Earlier it was thought that the set of absorption lines that we have called b must be the spectrum of the secondary component of the system. It was first described as of spectral type B2 and later as B5 and, therefore, β Lyrae was believed to be a B9 + B2e or a B9 + B5 system.

In 1934 Struve (1934) called attention to the abnormal line intensities that characterize the so-called "B2 or B5 spectrum", in which Curtis and also Baxandall had found practically no periodic variations. Struve suggested that those spectral lines were not coming from a stellar component but rather from "an expanding shell of gas". A few years later Struve (1941) further concluded that the shell of gas or large gaseous envelope extended to three times the stellar radius. The absorption line at HeI λ 3888 Å has multiple components which yield mean

velocities of -170, -135, -100 km sec^{-1} and there are probably other
components with smaller velocities (Sahade et al., 1959). The strongest component is the more violet-displaced. There are phase-dependent
variations in the appearance of the structure which are related with
the density distribution in the envelope at different phase angles.
This had been already noticed by Struve and, as a result, Gill (1941)
made measurements of equivalent widths which pointed in the same direction.

The existence of a large expanding envelope lent support to the interpretation of the change of period in terms of loss of mass.

Let us consider set c of absorption lines. The violet satellites
have been observed in the phase interval 0.001P-0.06P, and the red

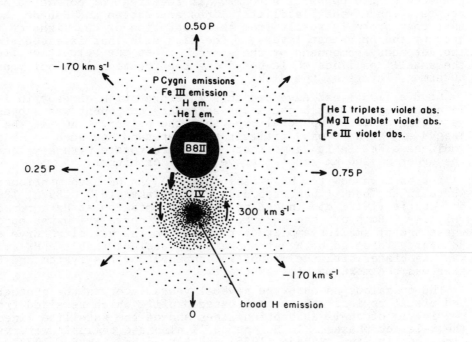

Fig. 10.2.1. Schematic model of β Lyrae.

satellites in the interval 0.92 P - 0.98 P. Both sets display strong
lines of H, HeI and of many other neutral and singly ionized elements,

but in the violet-displaced one, they are broad and even show a structure. The velocities are of the order of 250 km sec^{-1} on the mean.

In his 1941 paper Struve gave his interpretation of the satellite
lines as arising from two gaseous streams. The red satellites, from
a stream going from the secondary component towards the B8 companion,
and the violet satellites, from the stream going in the opposite direction and seen projected upon the disk of the B8 star after mid-of
primary eclipse. We have already described the changes that take place
in the light curve, in the spectral features and in the radial velocities immediately after mid-eclipse. Not only these changes but also
the strong effect of electron scattering that appears to be at work
immediately after the phase of minimum light at principal eclipse are
arguments that largely led Batten and Sahade (1973) to propose that
in β Lyrae there is only one stream that goes from the B8 component
towards the companion and that in the model of the system we should
omit the previously accepted opposite stream. Batten and Sahade's
suggestion is in better agreement with the observational facts, gives
a picture which is consistent with the results from other systems and
avoids the problem of understanding dynamically and evolutionary the
presence of two opposite streams. It requires, of course, a new interpretation of the "satellite" lines and Batten and Sahade thought
that they probably result from the absorption of the light of the B8
star by the outermost layers of the flat disk that is associated with
the secondary component of the system. Then the velocities from
the satellites lines would be giving the rotational velocity of the
outermost layers of the disk.

Struve (1934) had advanced this interpretation already in 1934 but
later he rejected it because the secondary component was supposed to
be a star of later spectral type than the primary - A0, at the earliest - and then it was difficult to understand how such an object
would have HeI in its atmosphere and would rotate with a velocity of
the order of 200 km sec^{-1} (Struve, 1941) or higher.

Let us turn now to the emission lines. As we have mentioned elsewhere in the book (Chapter 5.3.2), Belopolsky (1893, 1897a, 1897b)
and Curtis (1912) did indeed measure the emission at Hβ for radial
velocity. Both of them derived a velocity curve which was opposite in
phase and of smaller amplitude relative to the velocity curve of the
B8 star. This led both astronomers to conclude that the Hβ emission
was associated with the secondary component of the system and that this
star was the more massive of the two.

The emissions as observed several years later in the photographic
and visual regions had a complicated profile which resulted from the
cutting in of shell absorption lines and of the satellite lines at
the relevant phases. In Hα one could also see telluric watervapor
lines cutting in. Sahade (1966) and Batten and Sahade (1973) were
able to describe the emission profile at Hα by the combination of two
emission profiles, one broad and one relatively narrow (Chapter 5:3.
2; 4.2). The broad emission has a total width of about 1000 km sec^{-1}
and the relatively narrow one a total width of about 600-700 km sec^{-1}
and a central intensity several times that of the broad component.
The two emissions have different Balmer decrements. While the narrow
profile has a steep decrement, the broad one behaves in a more normal
fashion. This suggest that the narrow emission comes from a less
dense envelope than the broad profile. It was then concluded that the
broad emission probably arises in the envelope around the secondary

component of the system while the narrow emission originates in the outer envelope that surrounds the binary. That the emissions in β Lyrae originate in two different regions was confirmed by the analysis of the ultraviolet spectra taken with the TD1-A satellite (Hack et al., 1976a); such results were mentioned in Chapter 5 (3.2; 4.2).

High time-resolution spectroscopy has been applied to β Lyrae at Hα (Sanyal, 1976) and have disclosed phase-dependent variations in emission strength relative to the continuum in a time scale of a quarter of an hour to several hours. They can apparently be explained in terms of "shock waves generated by the streams impinging on the ring around the secondary".

As for the emissions with P Cygni profiles that are seen in the far ultraviolet region (from about 2250 Å shortwards), the absorption edges suggest velocities that agree with the most violet-displaced component of HeI λ 3888 Å and therefore, they must originate in the outer envelope.

β Lyrae has been extensively observed in the ultraviolet with the OAO-2, the Copernicus and the TD1-A satellites and from the Skylab, and several investigations have already been published (Hack, 1974; Hack et al., 1974a, 1974b, 1974c, 1976a, 1976B; Kondo et. al., 1976; Bless et al., 1976) and the results can be summarized as follows:

a)   in the interval λλ 1220-1800 Å the continuum, as well as the strong emission lines, exhibits large cycle-to-cycle variations;

b)   the continuum up to about λ 2300 Å has the energy distribution of the B8 II star;

c)   from about λ 2250 Å towards longer wavelengths an absorption spectrum is dominant, the features arising in the B8 component;

d)   from about λ 2250 Å towards shorter wavelengths the spectrum is dominated by strong emissions of low-excitation, many of them displaying P Cygni profiles that behave as indicated above;

e)   the far ultraviolet displays low excitation emissions of multi-ionized atoms with ionization potentials up to 97 eV, like NV;

f)   ultraviolet objective prism spectra taken with the Skylab and covering the region λλ 1400-2300 Å show many emissions some of them "probably due to intercombination transitions" like CIII λ 1909.

The observations and several lines of arguments suggest that turbulent mass motions on a massive scale are present in the envelope of β Lyrae and that the energy levels are collisionally excited.

We should discuss next the questions of the mass and the nature of the secondary component of the system and the source of excitation of the outer envelope.

Regarding the mass, we have mentioned the earlier results by Belopolsky and by Curtis, which suggested that the secondary component was the more massive of the two stars in the system. Then Kuiper arguing on the basis of the mass-luminosity relation and on the fact that the secondary spectrum is not detected, reached the conclusion that the masses were of the order of $78\mathfrak{M}_\odot$ for the primary and $52\mathfrak{M}_\odot$ for the secondary. This result was disturbing because it implied that the luminosity of the primary component, had to be over four magnitudes brighter than the value indicated by the spectrum. The alternative was clear; the secondary component had to be more massive than the primary.

About fifteen years after the Yerkes 1941 papers the trend began
to reverse. Gaposchkin (1956) went back to the old result of the mass
of the secondary being the largest and suggested that the object was
"camouflaged" by a thick atmosphere.

Then Sahade (1958) and Sahade et al. (1959), in the light of the
ideas on stellar evolution of single stars, stressed the concept that
the secondary component was the more massive, that its spectrum is an
emission line spectrum and that it might be an evolved star now in the
stage of being underluminous for its mass. Abt et al.(1962) analyzed
a few years later the multiple star system to which β Lyrae belongs
and derived a distance which gave for β Lyrae a visual luminosity of
-3.9, in fair agreement with Struve's earlier estimate from the spectrum. Now it is a general accepted fact that β Lyrae is a system where
the secondary component is more massive than the primary.

What is the nature of the secondary component? There have been many
speculations to answer this question. Two relatively recent papers
should be mentioned in this context. Wilson (1974) can explain β
Lyrae on the hypothesis of a thick disk, actually a flattened secondary, most of whose luminosity emerges at the poles. This region would
be 2000-3000°K hotter than the primary component and Wilson believes
that the radiation from the polar regions are the source of excitation
of the outer envelope. On the other hand, Kriz (1974) suggests a
"semitransparent scattering disk" around the secondary which would be
an apparently underluminous B0-B2 V star.

In this context, the infrared light curves of β Lyrae at 1.2, 2.2,
3.6, 4.8, and 8.6 μ obtained by Jameson and Longmore (1976) show that
the secondary minimum becomes deeper than the primary at longer wavelengths. Jameson and Longmore can explain the observations by postulating a plasma cloud around the secondary component.

Sahade (1976a) has suggested that perhaps our understanding of the
nature of the secondary component in β Lyrae will only come through
an analysis of the envelopes that are in the system. With the same idea
in mind, recently, Huang and Brown (1976) have analyzed the disk by
means of the light curves in different colours. Their conclusions
are that the secondary has to be an early B star if on the main sequence or an even earlier star if below the main sequence.

We are still not certain about the evolutionary history and stage
of β Lyrae. All possibilities have been proposed and we believe that
it is better not to refer to them. Let us just state that is one of
the many problems that need an answer.

About two centuries after its discovery β Lyrae is still one of the
more interesting objects in the sky, Fig. 10.2.1 sketches the model
of β Lyrae as suggested by our present knowledge of the system.

                          3. U CEPHEI

The system U Cephei has a relatively long history in the literature,
and is undergoing considerable activity at the time of this writing.
Its variability was discovered by Ceraski (1880) in June, 1880. At
the time, only six other eclipsing systems were known but their binary
nature had not been clearly established. Dugan (1920) lists two pre-
discovery observations, apparently in primary minimum in 1828 and 1855.

The spectral types, especially that of the fainter component, have been a matter of considerable discussion. After careful investigation, Batten (1974) finds that the system consists of a B7V and a G8III-IV component; the period is 2.5 days. The visual depth of primary is about 2.2 mag. and the depth of secondary only a few hundredths of a magnitude. A brief glance at the light curve suggests it is another typical Algol-type system showing a total eclipse. More careful study however presents quite a different picture.

Historically, the two chief puzzles have been (1) the observed changes of shape of the light curve during primary and (2) the discrepancy between the eccentricity of the orbit computed from the radial velocity curve and that indicated by the light curve. This latter was the first demonstrated by the large eccentricity shown by the velocity curve observed by Carpenter (1930) as contrasted to the central location of secondary in the light curve observed by Dugan (1920). This second difficulty is now largely resolved, since for more than twenty years we have been increasingly aware that in many systems the shape of the velocity curve is frequently distorted by circumstellar material and that eccentricities derived by conventional treatment in many cases are not reliable. U Cephei is almost certainly one of these.

The observed changes in the light curve have proven more difficult. Changes have been found both in and out of eclipse. One major difficulty has been that the brightness during the "constant phase" in totality, when the smaller component is entirely eclipsed and we see only the secondary, has at times indeed been constant, but at others has shown distinct variability. On the basis of reliable observations, there has been at some epochs a gradual increase of light during this interval and at others a gradual decrease. At still other times, "humps" or "dips" have been reported. At times these have exceeded 0.1 mag. in amplitude.

The shape in totality was discussed as long ago as 1889 (Chandler, 1889). Twenty years later Blazko (1909) suggested intrinsic variability of the large, cooler component. Dugan (1920) published a light curve obtained with a polarizing photometer and considered the possibility of a tidal bulge on the hotter component to explain the asymmetric shape of primary. By this time, the period change was recognized and the possibility of this being caused by the postulated bulge was discussed. Struve (1944, 1949) confirmed the distortion of the velocity curve and suggested a gaseous steam. Hardie (1950) attempted to correct for the blending of the lines and to derive the true orbital elements. In 1948, Walter (1948) reported a decrease in light during totality; by this time, the possibility of circumstellar matter was recognized. Walter considered this a more likely cause of the observed variation than intrinsic variability of the secondary itself. Later photoelectric observations by Miczaikza (1953) suggested to him an additional source of light on or near the B component. Bolokedze (1956) suggested that the asymmetry in primary was caused by the stream of gas which Struve had first suggested. Changes of depth of primary have also been reported (e.g., Bakos and Tremku, 1973), but it is difficult to make a historical study of these, because of the different color sensitivities of the equipment used by different observers.

The sudden period changes could be listed as another problem. However the general nature of these is similar to that found in certain other systems. Such changes were discussed in Chapter 6. A detailed period study of U Cephei has recently been published by Hall (1975) and by Hall and Keel (1977). The latter predict "with 65% confidence" a sudden period decrease between 1977 and 1981.

Just before the 1974 "outbreak", two extremely significant studies
of U Cephei appeared. One was by Batten (1974) and the other, based
almost entirely on analysis of previous photometric studies, was by
Hall and Walter (1974). Batten's paper gives a thorough review of
previous work on the system and presents a unified interpretation.
He find absolute dimensions, $R_B = 2.9 R_\odot$, $R_G = 4.7 R_\odot$, $Mg - 4.2 \mathfrak{M}_\odot$
$Mg = 2.8 \mathfrak{M}_\odot$

The evidence suggests a flattened disk rotating about the B star;
this should have about 10^{13} free electrons per cubic centimeter and
extend to about 2.8 of the B star's radius from the center of this
star. Mass transfer from the G to the B star would account for the
formation of the disk and also for the steady increase in orbital period. However as Batten points out, recent polarization measures by
Coyne (1974) cast some doubt on the existence of a permanent circumstellar disk, since they showed that polarization during eclipse was
small, constant, and "probably interstellar in origin". Later Piirula
(1975) found changing polarization during some eclipses. The changes
were not similar and were sometimes very small, indicating that the
presence of circumstellar matter may not be a permanent condition.
In words more prophetic than he could have envisioned, Batten stated,
"Continued study of the system is clearly desirable".

Hall and Walter treated U Cephei from two different aspects. First
they analyzed three different photoelectric light curves; Tschudovitchev
(1950), Khozov and Minaev (1969), and an unpublished one by Catalano
and Rodono. They were able to find one set of photometric elements
which satisfied all three. They found evidence (on all curves) of two
"hot spots" on the B-component; according to their analysis, one was
near a pole and the other near its equator. All three light curves
show evidence of absorption from a gas stream. This is evident preceding and during the descending branch of primary and, more weakly,
around fourth contact just as the partial phases of the eclipse are
ending. Hall and Walter feel confident in the elements they have derived (i = $82°.14$, $R_B = 0.1665$, $R_G - .3340$) since they satisfy light
curves at three different epochs and at three different effective wavelengths. The values found by Dugan from the visual observations were
i = $86°.4$, $R_B = 0.2000$, $R_G = 0.3223$; other values in the literature
fall in this same general range. The system is semidetached.

The second approach was to study 104 different light curves obtained
between 1880 and 1970. (This is in accord with Batten's first law and
also a remark we believe was first made by G. E. Kron - "The more a
system has been observed, the more it deserves to be observed".) They
defined a quantity, ΔM, which depends on the difference of brightness
at the beginning and the end of totality, in such a way that the light
during totality increases - "slants up" - when ΔM is positive and decreases when it is negative. These were arranged in chronological
order, assigned relative weights, and then normalized around 60 mean
epochs. Zero weight was given to four nights occurring almost exactly
at the epochs of sudden decrease of period.

A plot of the result shows relatively smooth and repetitive changes
of an irregular wave like form with a range of about 0.25 mag. between
the extremes. The curve is not periodic; the interval between succes-
ive maxima varies roughly between 10 and 15 years. With one exception
(which possibly can be explained on observational grounds), abrupt period decreases are correlated with positive values of ΔM and seem to

prefer the interval just preceding the maximum value of this quantity. Hall and Walter studied 17 other Algol systems with totalities of an hour or more and found a similar correlation. Thus, this may be a general characteristic of many close binaries at certain stages in their evolution.

The correlation between epochs of period decrease and ΔM can be explained by a combination of two effects. Hall and Garrison (1972) had suggested that the upward slant during totality could result from an accumulation of luminous matter which has been ejected from the subgiant, had passed around the B star, and has concentrated in front of its leading hemisphere. In the change of period mechanism proposed by Biermann and Hall (1973), the gas stream should have maximum strength during the time when the period is decreasing.

This has been a highly condensed summary of our state of knowledge of this system in 1974. Then, late in 1974, Olson (1974) found the duration of totality varied in different eclipses and occasionally in different wavelength bands in the same eclipse. Small (0.02 - 0.03 mag.) variations in brightness during totality also were observed.

This was closely followed by two contemporary papers (Plavec and Polidan, 1975; Batten et al, 1975) giving more details of the outburst. (It is interesting to note that these were both received by the same journal on the same day.) Plavec and Polidan reported very strong Hα emission during a primary eclipse on August 8, 1974. This was in contrast to preceding history; although Huang and Struve (1956) had predicted intermittent emission of U Cephei six previous efforts to find emission had failed to do so; only with great effort had Batten (1969) finally detected it on one night. Not only was strong emission observed during this 1974 eclipse, but marked changes were shown during the interval of observation (from phase 0.988 to 0.032). At the beginning of the total phase, the red displaced emission was "very strong" and the violet "marginal but real". As the eclipse progressed, the red grew weaker and the violet wing was clearly of greater intensity of the red comonent. This agrees with the well known model of a rotating ring as proposed by Joy for RW Tauri. Simultaneously with the spectrographic observations, L. McDonald has made photoelectric measures. He found the total phase shorter than that determined previously.

Almost exactly one month later, on September 7, 1974, Batten and colleagues observed very strong emission in all the Balmer lines from Hβ through H 18 in the H and K lines of Ca II, in λ 4481 Å of Mg II, and possibly in some lines of FeII and HeI. This suggested emission in a disk rotating around the primary and was seen in all observable eclipses in September and October, although it had noticeably weakened by 17 October. Photometric observations also showed changes in the shape of the light curve.

Actually, there had been earlier observations that unstable conditions were developing and quite possibly more than one outburst occurred. As early as 1969 Bakos and Tremko (1973) found that "in 20 days in August 1969, the phase shift increased by a factor of three over that in seven months between December 1968 and July 1969",..."the depth of minimum itself appears to be variable". Further, "It appears that the minima were deep and slightly asymmetrical in December 1968. Some two weeks later the depth has decreased by 0.1 mag. and by August 1969, the minimum became shallower by another 0.1 mag. At the same time the asymmetry has increased". In October, very weak emission

was observed during primary minima,   (Batten, 1969).   In 1970, the
careful spectrographic monitoring which had been carried out at the
Dominion Astrophysical Observatory at Victoria was temporarily dis-
continued.   However in September and October 1971, spectroscopic and
photometric observations at Victoria showed no emission and the light
curve was normal (A. H. Batten, personal communication).   In April
1972, Naftilan found much stronger emission than in 1969, although
this was not published until three years later.   In September and
October 1973, spectroscopic and photometric observations at Victoria
again showed everything normal (Batten, 1973) and polarimetric observations by Coyne (1974) also gave negative results.

The problem of when the major outburst really began is made difficult by the fact that changes in the light curves of U Cephei have
been observed so frequently.   However, things were apparently normal
in the autumn of 1973 and the outburst that followed exceeded by far
anything previously reported.

The light changes are now being studied more consistantly than at
any precious outburst.   As examples, we will summarize some of them.

Olson (1976a, 1976b, 1978) has discussed light changes both inside
and outside of primary minimum.   He monitored the system beginning
in January 1975 taking four color, ubvy and Hβ photoelectric observations; the yellow observations were transformed to the standard V
system.   During one cycle, narrow band observations centered Hα were
obtained.   The total number of approximately 10,000 observations obtained in this program are on file in the Variable Star Archives of
the Royal Astronomical Society, file no. IAU (27), RAS-48.   Changes
in primary minimum, especially in the duration of totality, are shown
quite clearly.   The following are some of the chief features shown
in the out-of-eclipse observations.

A dip near phase 0.6 which was developing on October 25, 1975 had
by October 30 reached a light loss of approximately 0.7 mag. in the
ultraviolet.   Observations on adjoining phases on nights near this
date showed extrapolations of it indicating a broad depression in the
light curve.   The appearance was apparently rather sudden, but nothing
is known about its activity after October 30.   Olson finds that broad
dips are often present during intervals of high activity and that they
can appear or disappear in a time scale of days.   They are not correlated with changes in Hα or Hβ and thus probably are continuum
effects.

A less prominent dip appeared about the same time around phase 0.2;
its maximum depth was about 0.4 mag. in the ultraviolet.   Variations
up to 0.3 mag. occurred between phases 0.35 and 0.51; the largest
losses seem to concide with periods of active mass transfer.   Smaller
"peaks" occured at other phases.   The smallest variations were found
near phase 0.88.   Olson concludes that during periods of high mass
transfer, the light curve departs "radically" from its "typical" state.

The model proposed to explain these changes follows earlier work
and suggests an equatorial disk of appreciable thickness around the
B star, a mass transferring stream, and a hot spot associated with the
B component.   The formation of large star spots on the photosphere of
the primary could explain the transient changes.   Two such spots are
required to explain the two major dips.   In this model, two parts of
the normal photosphere are each replaced by a spot of lower temperature assumed to be in local thermodynamic equilibrium.   A detailed
discussion shows that quantitative agreement with the observations,

such as the deficiency of ultraviolet light at some phases and an excess at others are not so easily explained.

Another extensive set of photometric observations has been obtained by N. L. Markworth (1977), who, at the date of this writing had obtained about 3200 photoelectric measures in the conventional UBV system. These covered the interval from 31 October 1974 through 19 May 1976.

Markworth also found the light changes between eclipses could not be treated by conventional methods of analysis. Two regions of the light curve seemed to be particularly troublesome. The observations on 15 October 1975 clearly show part of the dip reported by Olson at 0.6 P. Variations in primary minimum were noted, even on minima separated by only two orbital resolutions.

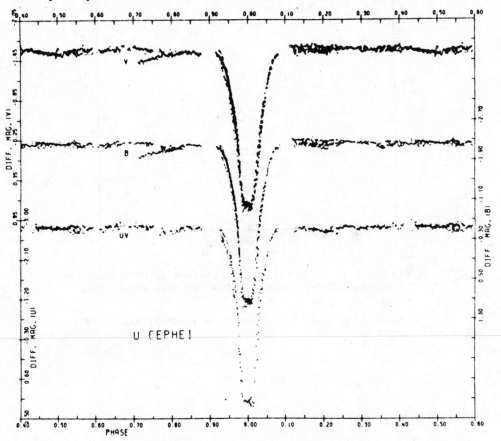

Fig. 10.3.1. U Cephei. Normal points. Observations by N. L. Markworth Rosemary Hill Observatory, October 1974 - May 1976.

In the region between phases 0.12 and 0.23, the observations fell into two distinct sets separated in brightness by 0.1 mag. The two sets were separated in time by approximately one year. The second region was that from phase 0.75 to primary eclipse. Here there was

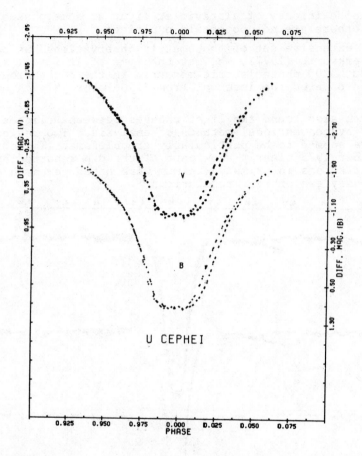

Fig. 10.3.2.   U Cephei minima observed on 10 November and
15 November 1974.   Comp.   * BD + 18° 29.   November is
"upper" curve on egress.

*Observation by N. L. Markworth at Rosemary Hill Observ-
atory.

a definite downward trend possibly indicative of effects of gas
stream.

Consideration of the models of gas streaming which took into account the hydrodynamics of such streams (e.g., Prendergast and Taam,
1974) suggests that the light level should be somewhat lower during
intervals of high activity; however, using only the higher level for
a Fourier analysis still did not give reasonable results, and neither

did application of the method used by Hall and Walter (1974); this
tried rectification using the only region supposedly free from gas
stream effects (from about 0.25 P to immediately preceding secondary).
Finally, by eliminating data which differed appreciably from the linear solution a reasonable fit was obtained, but it is clear that
conventional methods of treatment will not be satisfactory for U
Cephei, and presumably in other similar systems when observed during
or immediately following an outburst. Despite these difficulties,
Markworth was able to find a solution which agreed remarkably well
with that of Hall and Walter. He also developed a model for the system. Good quantitative results were obtained using a polar hot region.
Occasionally the observed eclipses seem to require in addition an
attenuating equatorial cloud of small optical depth. This model requires the primary component to rotate asynchronously (a well-known
conclusion from the radial velocity curve) in order to obtain good
fits to the observations. It will be interesting to follow future
work on this system.

4.   R CANIS MAJORIS

A number of close binary systems exist which show Algol-type light
curves and have periods of the order of a day or less. Their existence immediately presents an interesting problem. The short periods
indicate that the components must be extremely close, and yet no appreciable proximity effects are regularly repeated in the light curves.
The simplest explanation probably is the assumption of a relatively
large mass ratio. This means the more massive component, which contributes by far the major portion of the light, will be very little
distorted while the less massive, presumably a highly distorted component, contributes only a small fraction of the total light. In this
case, the light curve should be affected only in minor way and should
be that of the conventional Algol type.

R Canis Majoris (period about $1^d.2$) is one of the most thoroughly
studied of these systems. However, it would be a mistake to generalize and call it a representative of an entire group. It is true that
a "R Canis Majoris type" has been suggested and occasional references
to it are found in the literature. The assumption that the fainter
component had _precisely_ the same dimensions as its limiting lobe,
lead to the postulation of the existence of a group in which the components had masses appreciably smaller than one solar mass. However,
Sahade (1963) showed in discussing these systems that the method of
assuming one component must _exactly_ fill its lobe led to highly unreliable values for the masses, and further evidence leading to the
same conclusion was presented by Sahade and Ringuelet (1970).

More recently, Okazake (1977) has discussed T Leonis Minoris, originally listed as a member of this group, on the basis of UBV photoelectric light curves and intermediate dispersion spectra. In contrast
to the combined masses of the components of 0.84 $\mathfrak{M}_\odot$ (Kopal and Shapley,
1956) given when the system was listed in the "R Canis Majoris group",
he finds on this stronger evidence a combined mass of 3.27 $\mathfrak{M}_\odot$. In
discussing later photometric studies from various sources, Okazaki
also reaches the conclusion that these do not support the credibility
of the existence of such a group of systems.

From another point of view, in their discussion of anomalies in
light curves of eclipsing variables, Piotrowski et al. (1974) divide

the systems discussed into six categories; R Canis Majoris is the only
system in its category. Thus we discuss it here as an interesting
system in its own right and not as a representative of any particular
group or class.

The variability of this system was discovered by Sawyer in 1887
(Sawyer, 1887). The first unusual feature is shown in observations
made in 1896-99 (Pickering, 1904) and in 1898-99 (Wendell, 1909). This
consists of a "hump" in the light curve immediately following emergence
from primary minimum. These observations have been discussed by Dugan
(1924) and by Wood (1957) who conclude that was a real but non-perma-
nent feature of the light curve.

While there were many early series of observations, especially of
times of minima, the next complete and reliable light curve was obtained by Dugan (1924). Again a polarizing photometer was used; normally, different observers using this instrument have reported remarkably similar results. While the light curve resembled in many ways
that of Wendell, three features warranted special comment. (1) The
observed times of minimum now differed from those predicted and suggested a rather abrupt shortening of the orbital period in the interval between the two sents of observations. (2) No trace was found of
a "hump" on the light curve, although the appropriate phase interval
was well covered. (3) The depths of both primary and secondary minima were smaller in Dugan's observations. After a thorough discussion
of the data, Dugan concluded that the discrepancy in depths was the
outstanding difficulty.

The system was also observed spectrographically. Jordon (1916) presented a set of velocity curves taken in the years, 1908-12. The
system showed a single-lined spectrum. When the separate velocity
curves for the different years were solved in the conventional manner,
all showed large orbital eccentricities, although secondary eclipse
was centrally located in both light curves: A possible, although far-
fetched, explanation invoking a twenty year rotation of the apse with
the major axis pointed toward the observer at the epoch of each set
of photometric observations was ruled out by the fact that the observed minima over this interval showed no periodic fluctuation. Further, the velocity curves themselves were inconsistent; different
values of the eccentricity were computed if observations of different
years were used.

Today, in the light of our understanding of these systems as it
developed following the Struve revolution, a possible explanation presents no major problem. The humps observed by Wendell and by Pickering
could well have been associated with a developing instability resulting
in a violent mass ejection which caused the observed change of period.
If this ejected material had remained in the vicinity of the system
for the next twenty years, or if mass ejection had continued, the
circumstellar stream could well have distored the velocity curves and
have altered the depths of minima observed by Dugan. The amount of
extra light needed to force complete agreement in the observed depths
of minima is perhaps unconfortably large; 0.2 of the total light of
the system is required. This is, however, one case where "third light"
as discussed in Chapter 3 may be reasonable.

The next light curve by Wood (1946) was observed photoelectricly
in 1937-38 in the natural color system of the telescope-photometer
combination. When reduced to the wavelength of Wendell's observation,
this gave satisfactory agreement to the depths of minima found by him,

but disagreed with those found by Dugan. The light curve showed no
unusual features and suggested that, if such a thing is possible for
any eclipsing system, this binary had now returned to a "normal" state.

By no means did this imply that no problems remained connected with
the system. Since only one spectrum was visible, only the mass function could be derived directly and in order to compute absolute dimensions it was necessary to assume either a mass ratio or the mass
of one of the components. An interesting feature of R Canis Majoris
from the start was that fact that mass function derived for the system
was extremely small. As noted in Chapter 1, this posed quite a problem at a time when it was believed that mass ratios in spectroscopic
binaries should be close to unity. Any attempt to make the fainter
star "normal" in mass resulted in an absurdly large mass for the
brighter component, and attempts to give the brighter star a mass appropriate to its spectral class resulted in an extremely small mass
for the fainter object. However, some limits were set by use of the
concept of zero-velocity limiting surface, which of course depends on
the mass ratio, and by the assumption that the fainter component could
not exceed it - an assumption which later became commonly used. Even
this called for a mass ratio of about 5:1 or 6:1 and a mass for the
fainter component of less than $0.1 \, \mathfrak{M}_\odot$. However, the very small mass
function permitted no further alternatives at that time.

Later work includes confirmation of the very small mass function
by Struve (1948) and a radial velocity curve by Struve and Smith (1950)
which showed scattering and irregularities which have been attributed
to surrounding tenuous, gaseous matter. Fringant (1956) suggested a
spectral class of the brighter component as FIV and this was in per-
fect agreement with the (U-B), (B-V) colors observed by Kitamura (1969).
However, three color light curves by Koch (1960) and by Kitamura and
Takahashi (1962) suggested both components were appreciably overluminous. Interpretations were suggested by Smak (1961) and Sahade (1963).
Smak suggested the components were burning helium, and Sahade, the
possibility of the secondary overflowing the critical surface. More
recent work includes a radial velocity curve by Galeotti (1970) and
a three color light curve and extensive discussion by Sato (1971).

Sato discussed his V observations and compared them with those by
Kitamura and Takahashi, and by Koch. He found increased luminosity
in the middle of the interval between primary and secondary eclipses
in both of the later sets of observations. Those of Kitamura and
Takahashi also showed a "large depression" just before primary minimum and a "large asymmetry" in the light curve of the secondary eclipse. Sato's observations and those of Koch do not show such a large
assymmetry, but the former has a hump in the middle of secondary and
the latter, a small dip in the ascending branch.

Sato also observed the H_β index. This agreed with the FIV classification for the primary but showed variations of the same tendency
as variations of spectral lines being observed simultaneously by
Kitamura. Since his solution did not agree with the ones made from
the earlier observations, Sato used two methods to obtain the absolute dimensions. In the first, he assumed the fainter component was
in contact with the limiting surface, which gave the mass ratio as a
function of r_2. He then computed absolute dimensions for 6 different
sets of photometric elements and for four different values of f(m)
for each case - 24 sets in all. In the second approach, he repeated
these computations, assuming this time, however, that the primary had
a normal mass for its spectral type and taking types as A9V, F0V, F1V,

and F2V. On the basis of this, he concluded that R Canis Majoris is
an ordinary semi-detached system with a subgiant secondary of spectral
type G4 according to his effective temperature estimates. The observed peculiarities are caused by the gaseous envelopes. He concludes
that problems still remain - for example the discrepancy between the
absolute magnitude determined this way and that obtained from combining the apparent magnitude and the parallax.

Changes in the light curve are still being discussed. Piotrowski
and Rozyczka (1973) have reported possible physically significant
changes in the shape of primary minimum. Knipe (1963) had earlier
reported that a short constant phase had been found at the center of
primary minimum on two occasions but not in other minima observed
under equally favorable conditions. As mentioned earlier Piotrowski
et al.,(1974) have included R Canis Majoris in their studies of anomalies in light curves of eclipsing variables. The semi-detached systems were divided into three categories in the order of increasingly
stronger gas streams. R Canis Majoris is the only member of their
third (strongest gas stream) class, in which emission is very often
replaced by absorption.

Thus we find in R Canis Majoris a continuing series of minor problems of interpretation, although the general nature of the system now
seems clear. We note that this is not unusual when systems are near
enough to the solar system for their light and spectral changes to be
studied in detail and it seems reasonable to assume that all, or
nearly all, interacting binaries show similar effects. In R Canis
Majoris, probably the major unusual factor is the unusually small mass
of the secondary component.

5. ε AURIGAE

5.1. ε Aurigae is another one of the most interesting and puzzling
eclipsing systems. It has a period of 27.1 years - the longest so far
known among eclipsing variables - and its light curve is of the Algol-
type. Struve (Struve and Zebergs, 1962) has said that the history of
ε Aurigae "is in many respects the history of astrophysics since the
beginning of the 20th century".

The star was discovered as variable in 1821 by a German pastor named
Fritsch (1824), and the first observations that led to a knowledge of
the light curve were due to Schmidt, in Athens, and to Argelander, in
Bonn, who made visual estimates at the 1848 eclipse.

The early observing of the star was not confined to photometric
measurements but also included the securing of spectrographic material.
The latter began at about 1900 both at the Potsdam and at the Yerkes
Observatories.

It was Ludendorff (1903, 1924) who first determined the period and
ascertained the binary nature of the object with his publication of
an spectroscopic orbit.

5.2.1. Which are the other characteristics of the system? From the
point of view of the photometric behavior, we can say that the light
curve displays only one minimum. It is 0.8 magnitudes deep and remains at a more or less constant level during about a year while the
whole duration of the eclipse is nearly twice as long. Actually the

average durations of the "total" phase and of the entire eclipse, as
derived from the light observations made prior to 1956, are 330 days
and 714 days, respectively, while at the 1956 eclipse these figures
became 394 days and 670 days, respectively (Gyldenkerne, 1970)

The bottom of the eclipse, at least at the 1929 and 1956 minima,
has not been flat at all but did show humps of the order of 0.2 of a
magnitude (cf. Gyldenkerne, 1970); this is illustrated in Fig. 10.5.1.

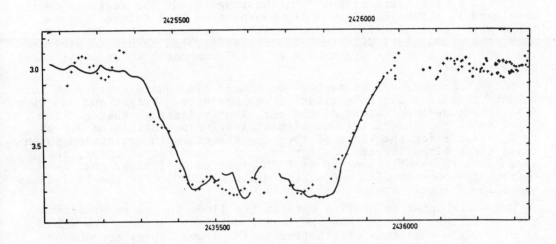

Fig. 10.5.1. The light curve of ε Aurigae at minimum as illustrated by
Gyldenkerne (1970). Smooth curve and dots, lower absicissa scale: 1955–57
eclipse; plus signs, upper abscissa scale: 1928–30 eclipse.

Light variations outside eclipse have also been reported by a number of observers. Stub (1972), for example, has used narrow-band
(about 80 Å half-width) filters peaked at nine wavelengths from λ 5520
He reports fluctuation at all wavelengths of amplitudes which decrease
with wavelength. The maximum value is about 0.2 mag observed in the
blue and ultraviolet. Stub also gives references to the earlier work
along the same lines.

Albo and Sorgsepp (1974) have reported a "flare" occurring early
in 1968. The duration from onset to finish was of the general order
of four days. The amplitude was approximately 0.2 mag in the UV with
smaller amplitudes in the B and V.

Coyne (1972) has reported small changes in polarization which are
possibly phase dependent. These consist of an increase following
phase 0.4. He plans to observe again after phase 0.6 to see if a
corresponding decrease is found.

5.2.2.   The spectrum of ε Aurigae is single-lined and corresponds to
an F2 I object (Hill et al., 1975). It is present at all phases even
during the long interval at which the light is at its lowest level,
while no features that could be assigned to the secondary component
has been ever observed.

As Struve (1956) has pointed out, the spectrum of ε Aurigae outside of eclipse clearly shows that the atmosphere of the F supergiant
component is very extended. The spectral features that arise in the
lower layers are weak, as is the case with the metallic lines, or are
broadened by the Stark effect. On the other hand, the features that
originate in the higher levels are sharp; thus, the Balmer lines -
the lowest level of which are metastable - are seen up to H29. Moreover, the geometrical dilution effect present, led Struve to estimate
that the extent of the atmosphere of the F component is at least three
times its radius.

In such an extended atmosphere we should find quite a range in turbulent velocities and in excitation temperatures. Wright and van Dien
(1949) have derived values of the turbulent velocities that go from
20 km sec^{-1} for the Fe I lines with excitation potentials of 0.5 eV
to 2 km sec^{-1} for the lines of the same element with excitation potentials eight times larger. As for the excitation temperatures, they
were found to be of the order of 6500° for the Fe I lines that have
metastable lower levels, and of the order of 10000° for the Fe I lines
whose lower levels are not metastable.

The differences in profile between the lines formed in different
regions further suggest that there is a velocity field in the atmosphere: while the lines originating in the lower layers are shaded
towards the red, those originating in the upper layers are shaded towards the violet.

Several strong metallic lines and Ca II and Na I display sharp violet edges that are characteristic of expanding shells and may suggest
that there is a large, expanding envelope around the whole system.

Hα, outside of eclipse, displays red and violet emission borders.
Their existence was reported for the first time by McKellar and Wright
(1957) who also found that "after first contact, an emission core at
approximately the normal wavelength appeared and a strong, broad absorption line developed, with, apparently, emission on the short wavelength side before mid-eclipse and on the long wavelength side after
mid-eclipse".

Wright and van Dien (1949) find that the outer edges of the emission components extend to 10-20 Å from the absorption core. They may
be formed in the large atmosphere of the F supergiant.

Long before the beginning of eclipse many lines do show a structure
and become normally double or even triple lines. A large number of
the extra lines belong to be same transitions as those that are present in the supergiant F star. They were first noticed by Adams and
Sanford (1930) and observed at the 1956 minimum by Struve at Mount
Wilson (Pillans, 1955; Struve, 1956; Struve and Pillans, 1957; Struve,
Pillans and Zebergs, 1958) and by Wright (1958) at Victoria, among
others. They appear at the red side - before mid-eclipse - or at the
violet side - after mid-eclipse - of the normal stellar features. At
mid-eclipse the intensities of the lines are stronger than outside of
eclipse. The separation of the redshifted components increases as
the eclipse progresses and reaches a maximum at conjunction, while

the separation of the violet-shifted component decreases from mid-eclipse towards the last contact.

Long before eclipse starts the effect of the extra features begins to be noticeable as an asymmetry in the stellar lines. Thus at the 1956 eclipse the unsymmetric profiles were already present on plates taken in 1951. Such an asymmetry in the lines is also present after the eclipse - although in the opposite direction - well past fourth contact.

Furthermore, during eclipse, some lines like Mg II λ 4481, Fe I λ 4260 and a few others are extremely weak, while the Cr I lines that originate from the ground level, are exceptionally strong. This behavior suggests the presence of diluted radiation.

We have mentioned that sometimes some lines appear triple during eclipse. Actually, in 1955 and 1956 Struve and Sahade secured a few spectrograms of ε Aurigae which clearly showed that at the very end of totality and a month later the stronger lines of Fe II and the strongest line of Ti II were triple while the rest of the elements, Sr II, Sc II, Ca I, Fe I, etc, appeared double. The violet component of the lines that appeared double were probably blends of two unresolved components. This is suggested by the radial velocities derived from the components in the case of the "double" and of the "triple" lines.

5.3. The radial velocities of ε Aurigae that are available have been combined (Morris, 1962; Wright, 1970) to derive the orbital elements that are transcribed in Table 10.5.2.

Table 10.5.2.

ORBITAL ELEMENTS OF ε AURIGAE

K	15.0	km sec^{-1}
γ	$-$ 1.4	km sec^{-1}
e	0.20	
a_1 sini	2.0×10^{9}	km
f($\mathfrak{M}$)	3.12	$\mathfrak{M}_{\odot}$

The period preferred by Gyldenkerne (1970) is 9885 days, which was derived by comparing two photoelectrically observed minima.

The actual radial velocities at different epochs show departures or fluctuations that may be as large as 20 km sec^{-1} relative to the computed velocity curve.

5.4.   Let us go now to the problems that are posed by the system.
In the first place we have, roughly speaking, a flat-bottomed Algol-
type light curve. This normally means that at primary eclipse the
brighter component is eclipsed by a larger, dark companion. However,
the spectrum of the F2 I star is seen at all phases, even when the
drop in light has reached its maximum value; consequently, such an
explanation does not seem to hold in this case.

Furthermore, the drop in light is of the order of 0.8 magnitude
and this would indicate that the two components have about the same
surface brightness; hence, we should have two minima of similar depths
However, not only is there no secondary minimum with a depth of the
same order as the depth of the primary eclipse but, in fact, no secondary minimum has been detected at all. There can be no doubt about
this because the orbital period suggested by the radial velocities is
precisely the time interval between two consecutive minima. Gaposchki
(1954) tried to solve the puzzle by assuming that we are actually
dealing with an annular eclipse of the F2 star and that this component
was "edgeless", that is, had no limb darkening. The possibility was
somewhat attractive, but the proposed situation cannot produce the
observed depth of the eclipse.

What is then the nature of the secondary component? Kuiper, Struve
and Strömgren (1937) were the first to try to answer this question.
They proposed that the companion to the F supergiant was a large, semi
transparent infrared star, which was called the "I" star and had a
radius of 3000 times the solar radius. They further proposed that a
thin layer of the hemisphere of the "I" object that faces the companion, was ionized by the latter's ultraviolet radiation. Electron
scattering in the ionosphere-like layer was then responsible for the
nearly constant opacity of the large body throughout the "total" phase
of minimum light. Electron scattering alone, however, does not seem
to be sufficient to reproduce the observed depth of minimum and, moreover, it poses other problems. Furthermore, an infrared component
with the characteristics suggested in the proposed model cannot be
understood in terms of internal structure and in terms of the observational facts during the eclipse phases. We could add that, on the
other hand, the results of the infrared photometry of ε Aurigae at
8 μ and 9.2 μ (Low and Mitchell, 1964) do not provide any evidence
for the existence of an infrared component in the system. As a consequence, other sources of opacity have been suggested, namely, fairly
large solid particles (Schönberg and Jung, 1938) and H⁻ (Strömgren:
cf. Kraft, 1954).

The need to explain the line asymmetries and extra absoprtions that
appear at and near light minimum led Struve to a few suggestions as
to their origin. In 1955 (cf. Struve, 1962) he thought, without drop-
ping the idea of the opacity being the result of electron scattering,
of a circumstellar envelope around the secondary that would be broken
up in a number of separate "clouds". The velocities of these clouds
would be distributed in an irregular manner but they would fundament-
ally move in the direction of the orbital motion (Struve and Pillans,
1957).

Slowly the models of ε Aurigae with a secondary component larger
than the primary were abandoned and new ones with a small star in
front at eclipse began to be put forward. From an anlysis of the
dilution effect, Hack (1961) concluded that the eclipsing star must
be a small, hot object. Earlier Sahade (1958) had suggested the possibility of a small, massive, underluminous secondary. The problem

of explaining the "total" phase was still open a few years later. Huang (1965) proposed that the small secondary was surrounded by a flat, rotating, gaseous disk that appears opaque when viewed edge-on. Huang's model is similar to the one he had advanced earlier to account for the case of β Lyrae and his idea finds an antecedent in Kopal's (1954) suggestion of a semitransparent flat ring of solid particles surrounding the secondary component.

Huang's model met some criticism from Wilson (1971) who came with the proposal of a thin opaque disk of dust around the secondary, the disk having a semitransparent central opening. More recently, Handbury and Williams (1976) have considered a disk of dust and gas, which is opaque across its center and becomes transparent near the edges, much as we believe it is so in the case of β Lyrae (see section 2 of this Chapter. In Handbury and Williams' view the primary component would be contracting towards the main sequence.

The most recent suggestion is due to Paczyński (1975) who evisages ε Aurigae as a sort of Algol-type system where the F supergiant fills its critical equipotential lobe, and the stream of matter that is generated leads to an accretion disk around the companion.

The spectroscopic behavior of ε Aurigae at and around conjunction is typical of an atmospheric eclipse, although certainly a more complex one in our case than in those that we discussed in section 7 of Chapter 9.

Taking into account the conclusions from other systems and the different suggestions and criticisms made in the case of ε Aurigae we feel inclined to describe the structure around the secondary component of the latter system in terms of Handbury and Williams' suggestion of a flat disk. It is certainly possible to think that in addition to the flat disk there is a general tenuous circumstellar envelope around the star and that such a structure is broken up in clumps. However, at least part of the story that the extra lines are telling us may be related to the rotation of the flat disk. Electron scattering appears to be the likely source of opacity and we have mentioned evidence for the existence of a large expanding envelope that surrounds the whole system.

Roughly, the model that seems to emerge does not differ very much from the models that we have described elsewhere in the book, particularly to explain β Lyrae. However, the spectral and photometric behavior of ε Aurigae is still not completely understood and more observations are required. Perhaps the forthcoming eclipse of the system will provide the opportunity for gathering the much needed additional information.

FINAL COMMENTS

The preceding chapters are by no means an exhaustive treatment of the present status of the field in its different facets. We have only dealt with the parts of the subject that seem to us the most important ones at the present time to introduce researchers to the field and to give necessary background to scientists who come from other fields of science. Almost any of the chapters could itself easily be expanded to a book.

In no part of the book we have given the usual classification of close binaries. It is well known that from the point of view of the light behavior eclipsing binaries are divided in Algol, β Lyrae, and W Ursae Majoris systems. However with our present knowledge on the subject we feel that the use of these denominations without a clear definition of precisely what is meant is misleading. An Algol-type light curve means a light curve where the light between minima is practically constant, while a light curve where the light variation is continuous is called a β Lyrae or a W Ursae Majoris type, depending on whether the period is larger or shorter than one day, on whether or not the minima are of approximately equal depth, and, in the case of the latter group, if the components are late type objects.

On the other hand, an Algol-type system is a system that is formed by an A-F main sequence primary and a late type subgiant secondary. And then we run into the contradiction of having systems with "Algol-type light curve" which are not "Algol-type systems". Furthermore β Lyrae, with all its complexities, is practically a unique system. The W Ursae Majoris systems are, as far as we know today, a definite group of objects and, therefore, there is nothing wrong with keeping the present designation for both the light curve and the object themselves.

In the other cases we would like to suggest that we drop the designation of Algol or β Lyrae for the light curves and use instead some designations such as "flat" and "rounded" eclipsing light curves, respectively. However, finding an adequate scheme presents difficulties. One has only to look through "An Atlas of Light Curves of Eclipsing Variables" compiled by Mario G. Fracastoro, to realize the complexity of the task.

We close with one final remark. A few years ago, a distinguished astronomical colleague returned to his home observatory from a visit to a country where some rather dramatic events had taken place. With the rather skeptical attitude of true scientists, his fellow astronomers asked him whether all the reports were true. He assured them they were and gave graphic descriptions. "It must have been very exciting", they said. "Well, yes" was the reply, "but not nearly as exciting as close binary systems".

We concur.

BIBLIOGRAPHY

Ables, J. G. (1969) Astrophys. J. (Letters) 155, L27.
Abt, H. A., Jeffers, H. M., Bitson, J., and Sandage, A. R. (1962)
     Astrophys. J. 135, 429.
Acker, A. (1976) Publ. Obs. astr. Strasbourg 4, fasc. 1.
Adams, W. S. and Sanford, R. F. (1930) Publ. astr. Soc. Pacific
     42, 203.
Ahnert, P. (1974) Mitt. veränderl. Sterne 6, 143.
Albo, H. and Sorgsepo (1974) Publ. Tartu Astrophys. Obs. 42, 166.
Allen, D. A. and Porter, F. C. (1973) Astr. Astrophys. 22, 159.
Allen, D. A., Swings, J. P., and Harvey, P. M. (1972) Astr.
     Astrophys. 20, 333.
Aller, L. H. (1977) J. R. astr. Soc. Can. 71, 67.
Altenhoff, W. J., Braes, L. L. E., Habling, H. J., Olnon, F. M.,
     Schoemaker, A. A., van den Heuvel, E. P. J., and Wendker, H. J.
     (1973) IAU Circ. No. 2549.
Altenhoff, W. J., Braes, L. E. E., Olnon, F. M., and Wenker, H. J.
     (1976) Astr. Astrophys. 46, 11.
Andrew, B. H. and Purton, C. R. (1968) Nature 218, 855.
Andrews, P. J. (1967) Astrophys. J. 147, 1183.
Apparao, K. M. V. and Chitre, S. M. (1976) Space Sci. Reviews.
     19, 28.
Appenzeller, I. and Hiltner, W. A. (1967) Astrophys. J. 149, 353.
Arnold, C. N. and Hall, D. S. (1973) IAU Inf. Bull. Var. Stars
     No. 843.
Atkins, H. L. and Hall, D. S. (1972) Publ. astr. Soc. Pacific
     84, 638.
Avni, Y. (1976) Astrophys. J. 209, 574.
Babcock, H. W. (1953) Mon. Not. R. astr. Soc. 113, 368.
Babcock, H. W. (1958) Astrophys. J. Suppl. Ser. 3, 201.
Bachmann, P. S. and Hershey, J. L. (1975) Astr. J. 80, 836.
Bakus, G. A. and Tremko, J. (1973) Bull. astr. Inst. Csl. 24, 298.
Bappu, M. K. V. and Sahade, J. (1973) Wolf-Rayet and High-Temper-
     ature Stars., ed. Bappu, M. K. V. and Sahade, J., p. 222.
     D. Reidel, Dordrecht.
Bappu, M. K. V. (1973) Wolf-Rayet and High-Temperature Stars, eds.
     Bappu, M. K. V. and Sahade, J., p. 54, D. Reidel, Dordrecht.
Barr, J. M. (1908) J. R. astr. Soc. Can. 2, 70.

Bath, G. T. (1972) Astrophys. J. _173_, 121.
Bath, G. T., Evans, W. D., Papaloizou, J., and Pringle, J. E. (1974)
    Mon. Not. R. astr. Soc. _169_, 447.
Bath, G. T. (1977) Mon. Not. R. astr. Soc. _178_, 203.
Batten, A. H. (1967) Publ. Dom. astrophys. Obs., Victoria _13_, 8.
Batten, A. H. (1968) Astr. J. _73_, 551.
Batten, A. H. (1969) Publ. astr. Soc. Pacific _81_, 904.
Batten, A. H. (1970) Publ. astr. Soc. Pacific _82_, 574.
Batten, A. H. (1970) Mass Loss and Evolution in Close Binaries, eds.
    K. Gyldenkerne and R. M. West, p. 101 Copenhagen Univ. Publ.
    Fund.
Batten, A. H. (1973) Personal communication.
Batten, A. H. (1973a) Binary and Multiple Systems of Stars, p. 196,
    Pergamon Press.
Batten, A. H. (1973b) Extended Atmospheres and Circumstellar Matter
    in Spectroscopic Binary Systems, ed. A. H. Batten, p. 1 D.
    Reidel, Dordrecht.
Batten, A. H. (1974) Dominion Astrophys. Obs. _14_, 191.
Batten, A. H., Fisher, W. A., Baldwin, B. W., and Scarfe, C. D.
    (1975) Nature _253_, 174.
Batten, A. H., and Laskarides, P. G. Publ. astr. Soc. Pacific _81_,
    677.
Batten, A. H., and Sahade, J. (1973) Publ. astr. Soc. Pacific _85_,
    599.
Baxandall, F. E. (1930) Ann. Solar Phys. Obs. Cambridge 2, part 1.
Beals, C. S. (1968) Wolf-Rayet Stars (NBS SP-307), ed. K. B. Gebbie
    and R. N. Thomas, p. 7.
Belopolsky, A. (1893) Mem. Soc. Spett. Ital. _22_, 101.
Belopolsky, A. (1897a) Mem. Soc. Spett. Ital. _26_, 135.
Belopolsky, A. (1897b) Astrophys. J. _6_, 328.
Benson, R. S. (1970) Ph.D. Dissertation, Univ. of California,
    Berkeley.
Bidleman, W. P. (1954) Astrophys. J. Suppl. _1_, 218.
Biermann, P. (1971) Astr. Astrophys. _10_, 205.
Biermann, P. and Hall, D. S. (1973) Astr. Astrophys. _27_, 249.
Biermann, P. and Thomas, H-C (1973) Astr. Astrophys. _23_, 55.
Blanco, C. and Catalano, S. (1970) Astr. Astrophys. _4_, 482.
Blazko, S. (1909) Astr. Nachr. _181_, 295.
Bless, R. C., Eaton, J. A., and Meade, M. R. (1976) Publ. astr. Soc.
    Pacific _88_, 899.
Bloomer, R. H. and Wood, F. B. (1974) Publ. astr. Soc. Pacific _86_
    689.
Blumenthal, G. R. and Tucker, W. H. (1974) Ann. Rev. Astr. Astrophys.
    _12_, 23.
Bodenheimer, P. and Ostriker, J. P. (1970) Astrophys. J. _161_, 1101.
Bolton, C. T. (1972) J. R. astr. Soc. Can. _66_, 219 (abstract).
Bolton, C. T. (1973) Extended Atmospheres and Circumstellar Matter
    in Spectroscopic Binary Systems, ed. A. H. Batten, p. 71, D.
    Reidel, Dordrecht.
Bond, H. E. (1972) Publ. astr. Soc. Pacific _84_, 839.
Bopp, B. W. and Fekel, F. (1976) Astr. J. _81_, 771.
Borra, G. F. and Landstreet (1973) Astrophys. J. (Letters) _185_,
    L139.
Botsula, R. A. (1962) Bull. astr. Obs. V. P. Engel'gardta, No. 37.
Bottinger, K. F. (1926) Astr. Nachr. _226_, 239.
Boyarchuk, A. A. (1959) Soviet Astr. _3_, 748.
Boyarchuk, A. A. (1969) Non-Periodic Phenomena in Variable Stars,
    ed. L. Detre, p. 395, Academic Press.

Bradt, H. V., Braes, L. L. E., Forman, W., Hesser, J. E., Hiltner,
    W. A., Hjellming, R., Kellog, E., Kunkel, W. E., Miley, G. K.,
    Moore, G., Pel, J. W., Thomas, J., VandenBout, P., Wade, C., and
    Warner, B. (1975) Astrophys. J. 197, 443.
Bradt, H. V., Apparao, K. M. V., Dower, R., Donsey, R. E., Jernigan,
    J. G., and Markert, T. H. (1977a) preprint.
Bradt, H. V., Apparao, K. M. V., Clark, G. W., Dower, R., Donsey,
    R., Hearn, D. E., Jernigan, J. G., Joss, P. C., Mayer, W.,
    McClintock, J., and Walter, F. (1977b) preprint.
Braes, L. L. E. and Miley, G. K. (1971) Nature 232,246.
Braes, L. L. E. and Miley, G. K. (1972) Nature 237, 506.
Brown, R. H., Davis, J., Herbison-Evans, D., and Allen, L. R. (1970)
    Mon. Not. R. astr. Soc. 148, 103.
Brucato, R. J. and Zappala, R. R. (1974) Astrophys. J. (Letters)
    189, L71.
Budding, E. (1973) Astrophys. Space Sci. 22, 87.
Budding, E. (1974) Astrophys. Space Sci. 26, 371.
Budding, E. (1974) Astrophys. Space Sci. 29, 17.
Bulokedze, R. D. (1956) Var. Stars 11, 375.
Burbidge, G. R. (1967) Radio Astronomy and the Galactic System, ed.
    M. van Woerden, p. 463, Academic Press.
Cameron, A. G. W. (1971) Nature, 229, 178.
Campbell, W. W. (1901) Lick Obs. Bull. 1, 22.
Canizares, C. R., Clark, G. W., Li, F. K., Murthy, G. T., Bardas,
    D., Spritt, G. F., Spencer, J. H., Mook, D. E. Hiltner, W. A.,
    Williams, W. L., Moffett, T. J., Grupsmith, G., VandenBout, P.
    A., Golson, J. C., Irving, C., Frohlich, A., and van Genderen,
    S. M. (1975) Astrophys. J. 197, 457.
Carpenter, E. F. (1930) Astrophys. J. 72, 205.
Carruthers, G. P. (1968) Astrophys. J. 151, 269.
Catalano, S. and Rodonó, M. (1967) Mem. Soc. astr. Ital. 38, 395.
Catalano, S. and Rodonó, M. (1969) Non-Periodic Phenomena in
    Variable Stars, p. 435, ed. L. Detre, Academic Press, Budapest;
    Catania Publ. No. 134.
Catalano, S. and Rodonó, M. (1974) Publ. astr. Soc. Pacific 86, 390.
Cayrel de Stroble, G., Chalonge, D., and Divan, L. (1955) Mem.
    Soc. astr. Ital. 26, 257.
Ceraski, W. (1880) Astr. Nachr. 97, 319.
Ceraski, W. (1914) Astr. Nachr. 197, 256.
Chambliss, C. R. (1976) Publ. astr. Soc. Pacific 88, 762.
Chandler, S. C. (1899) Astr. J. 19, 49.
Chandler, S. C. (1901) Astr. J. 22, 39.
Chandrasekhar, S. (1933) Mon. Not. R. astr. Soc. 449, 462, 539.
Chen, K-Y, Merrill, J. E., and Richardson, W. W. (1977) Astr. J.
    82, 67.
Chen, K-Y and Rhein, W. J. (1969) Publ. astr. Soc. Pacific 81, 387.
Chen, K-Y and Rhein, W. J. (1971) Publ. astr. Soc. Pacific 83, 449.
Chen, K-Y and Rhein, W. J. (1973) Acta astr. 23, 247.
Chen, K-Y and Reuning, E. G. (1966) Astrophys. J. 90, 449.
Chen, K-Y and Wood, F. B. (1975) Astrophys. J. (Letters) 195, L73.
Chen, K-Y and Wood, F. B. (1976) Mon. Not. R. astr. Soc. 176, 5P.
Cheng, C. C., Phillips, K. J. H., and Wilson, A. M. (1974) Nature
    251, 589.
Cherapaschuk, A. M. (1969) Astr. Cirk. No. 509.
Cherapaschuk, A. M. (1975) Astr. Zu. 19, 47.
Christie, W. H. (1933) Publ. astr. Soc. Pacific 45, 258.
Christie, W. H. (1936) Astrophys. J. 83, 433.
Christy-Sackmann, J. and Despain, K. H. (1974) Astrophys. J. 189, 523.

Clark, T. A. Hutton, L. K., Ma, C., Shappiro, I. I., Wittels, J.
    J., Robertson, D. S., Hinteregger, H. F., Knight, C. A., Rogers,
    A. E. E., Whitney, A. R., Niell, A. E., Resch, G. M. and
    Webster, Jr., W. J. (1976) Astrophys. J. (Letters) 206, L107.
Clayton, D. D. (1968) Principles of Stellar Evolution and Nucleo-
    synthesis, p. 155, McGraw-Hill Book Co.
Conti, P. S. (1977) preprint.
Cowley, A. P. (1964) Astrophys. J. 139, 817.
Cowley, A. P. (1967) Astrophys. J. 147, 609.
Cowley, A. P. (1969) Publ. astr. Soc. Pacific 81, 297.
Cowley, A. P., Hiltner, W. A., and Berry, C. (1971) Astr. Astrophys.
    11, 407.
Cowling, T. G. (1938) Mon. Not. R. astr. Soc. 98, 734.
Coyne, G. V. (1972) Ric. astr. Specola astr. Vatic 8, 311.
Coyne, G. V. (1974) Ric. astr. Specola astr. Vatic 8, 475.
Crampton, D. (1974) Astrophys. J. 187, 345.
Crawford, J. A. (1955) Astrophys. J. 121, 71.
Crawford, J. A. and Kraft, R. P. (1956) Astrophys. J. 123, 44.
Crawford, J. A. and Kraft, R. P. (1965) Publ. astr. Soc. Pacific 67,
    387.
Cristaldi, S., Fracastoro, M. G., and Sobieski, S. (1966) Mem. Soc.
    astr. Ital. 37, 347.
Curtiss, R. H. (1908) Astrophys. J. 28, 150.
Curtiss, R. H. (1912) Publ.Allegheny Obs. 2, 73.
de Monteagudo, V. N. and Sahade, J. (1970) Observatory 90, 198.
de Monteagudo, V. N. and Sahade, J. (1971) Observatory 91, 220.
Deutsch, A. J. (1956) Astrophys. J. 123, 210.
Doherty, L. R., McNall, J. F., and Holm, A. V. (1974) Astrophys. J.
    187, 521.
dos Reis, L. C.,(1975) Private Communication.
dos Reis, L. C. (1976) Private Communication.
Dugan, R. S. (1908) Publ. astr. Astrophys. Soc. America 1, 311.
Dugan, R. S. (1909) Publ. astr. Astrophys. Soc. America 1, 320.
Dugan, R. S. (1911) Contr. Princeton Obs. No. 1.
Dugan, R. S. (1920) Contr. Princeton Obs. No. 5
Dugan, R. S. (1924) Contr. Princeton Obs. No. 6.
Dugan, R. S. and Wrights, F. W. (1937) Contr. Princeton Obs. 19.
Eaton, J. A. (1975) Publ. astr. Soc. Pacific 87, 745.
Ebbighausen, E. G. and Gange, J. J. (1963) Publ. Dom. Astrophys.
    Obs., Victoria 12, 5.
Eddington, A. S. (1926) Mon. Not. R. astr. Soc. 86, 320.
Eggen, O. J. (1961) R. Obs. Bull. No. 31.
Eggen, O. J. (1967) Mem. R. astr. Soc. 70, 111.
Elvey, C. T. and Babcock, H. W. (1943) Astrophys. J. 97, 412.
Ezer, D. (1961) Commun. Faculty Sci. Ankara Univ. Ser. A, 11, 40.
Faraggiana, R. A. and Hack, M. (1966) Z. Astrophys. 64, 48.
Faulkner, J. (1971) Astrophys. J. (Letters) 170, L99.
Faulkner, J. (1974) Late Stages of Stellar Evolution, ed. R. J.
    Taylor, p. 155, D. Reidel, Dordrecht.
Ferrari, K. (1934) Astr. Nachr. 253, 225.
Fetlaar, J. (1923) Rech. astr. Obs. Utrecht 9, No. 1.
Flannery, B. P. (1975) Astrophys. J. 201, 661.
Flannery, B. P. (1976) Eighth Texas Symp. on Relativistic Astrophys.
Fletcher, E. S. (1964) Astr. J. 69, 357.
Florkowski, D. R. (1975) Bull. Amer. astr. Soc. 7, 56.
Florkowski, D. R. (1977) Private Communication.
Florkowski, D. R. and Gottesman, S. T. (1976) Inf. Bull. Variable
    Stars No. 1101.

Florkowski, D. R. and Gottesman, S. T. (1977) Mon. Not. R. astr.
    Soc. 179, 105.
Fredrick. L. W. (1960) Astr. J. 65, 628.
Frieboes-Conde, H. and Herczeg, T. (1973) Astr. Astrophys. Suppl.
    12, 1.
Frieboes-Conde, H., Herczeg, T., and Høg, E. (1970) Astr. Astrophys.
    4, 78.
Friedjung, M., (1977) Novae and Related Stars, D. Reidel, Dordrecht.
Fringant, A. M. (1956) Contr. Inst. Astrophys. Paris Ser. A.,
    No. 216.
Fritsch, J. H. (1824) Berl. Jahrb, p. 252.
Galatola, A. (1972) Astrophys. J. 175, 809.
Galeotti, P. (1970) Astrophys. Space Sci. 7, 87.
Gaposchkin, S. (1937a) Publ. Amer. astr. Soc. 9, 39.
Gaposchkin, S. (1937b) Harvard Circ. No. 421.
Gaposchkin, S. (1938) Publ. Amer. astr. Soc. 9, 152.
Gaposchkin, S. (1940) Publ. Amer. astr. Soc. 10, 52.
Gaposchkin, S. (1941) Astrophys. J. 93, 202.
Gaposchkin, S. (1949) Perem. Zvezdy 7, 36.
Gaposchkin. S. (1954) Publ. astr. Soc. Pacific 66, 112.
Gaposchkin, S. (1956) Zeitschrift f. Astrophys. 39, 133.
Gibson, D. M. and Hjellming, R. M. (1974) Publ. astr. Soc. Pacific
    86, 652.
Gibson, D. M., Hjellming, R. M., and Owen, F. N. (1975a) Astrophys.
    J. (Letters) 200, L99.
Gibson, D. M., Hjellming, R. M., and Owen, F. N. (1975b) IAU Circ.
    No. 2789.
Gill, J. R. (1941) Astrophys. J. 93, 118.
Glasby, J. S. (1970) The Dwarf Novae, American Elsevier, New York.
Goedicke, V. (1939a) Observatory 62, 197.
Goedicke, V. (1939b) Publ. Obs. Univ. Michigan 8, 1.
Goodricke, J. (1783) Phil. Trans. R. Soc. London 73, 474.
Goodricke, J. (1785) Phil. Trans. R. Soc. London 75, 40, 153.
Gorenstein, P. and Tucker, W. H. (1976) Ann. Rev. Astr. Astrophys.
    14 373.
Gould, N. (1957) Publ. astr. Soc. Pacific 69, 541.
Gould, N. (1959) Astr. J. 64, 136.
Greenstein, J. L. (1940) Astrophys. J. 91, 438.
Greenstein, J. L. (1960) Stellar Atmospheres, p. 676, ed. J. L.
    Greenstein, Univ. Chicago.
Greenstein, J. L. (1973) Astr. Astrophys. 23, 1.
Greenstein, J. L. (1976) Mém. Soc. R. d. Sciences de Liége, 6éme
    série 9, 247.
Greenstein, J. L., Hernández, C., Milone, L., Sahade, J., and
    Thackeray, A. D. (1970) Inf. Bull. 5th Hemisphere No. 16, p. 40.
Greenstein, J. L. and Page, T. L. (1941) Astrophys. J. 93, 128.
Gregory, P. C., Kronberg, P. P., Seaquist, E. R., Hughes, V. A.,
    Woodsworth, A., Viner, M. R., and Retallack, D. (1972a) Nature
    239, 440.
Gregory, P. C., Kronberg, P. P., Seaquist, E. R., Hughes, V. A.,
    Woodsworth, A., Viner, M. R., Retallack, D., Hjellming, R. M.,
    and Balick, B. (1972b) Nature Phys. Sci. 239, 114.
Gregory, P. C., Kwok, S., and Seaquist, E. R. (1977) Astrophys. J.
    211, 429.
Grygar, J. (1962) Kleine Veröff. Remeis-Sternw. 34, 26.
Grygar, J. and Horak, T. B. (1974) Bull. astr. Inst. Csl. 25, 275.
Guinan, E. F. and McCook, G. P. (1974) Publ. astr. Soc. Pacific
    86, 947.

Guinan, E. F., McCook, G. P., Bachmann, P. J., and Bistling, W. G.
    (1976) Astr. J. 81, 57.
Günther, O. (1955) Mitt. de Astr. Ges. 1954, p. 23
Günther, O. (1959) Astr. Nachr. 285, 97.
Gursky, H. (1976) Structure and Evolution of Close Binary Systems,
    p. 19, ed. P. Eggleton et al., D. Reidell, Dordrecht, IAU Symp.
    No. 73.
Gursky, H. and Ruffini, R. (1975) Neutron Stars, Black Holes and
    Binary X-ray Sources, D. Reidel, Dordrecht.
Gursky, H and Schreier, E. (1975) Variable Stars and Stellar Evolu-
    tion, p. 413, ed. V. E. Sherwood and L. Plaut, D. Reidel,
    Dordrecht.
Guseinov, O. Kh. and Novruzova, Kh. I. (1974) Astrofizika 10(2),
    273. [trans. Astrophys. 10, 163 (1975)].
Gusseinzade, A. A. (1969) Perem. Zvezdy 16, 500.
Güssow, M. (1936) Veröff. Berlin Babelsberg 11, 1.
Guthnick, P. and Schneller, H. (1932) Sber. prus. Akad. Wiss. f.
    1932, p. 59.
Gyldenkerne, K. (1970) Vistas in Astronomy, Vol. 12, p. 199, ed. A.
    Beer, Pergamon Press, Oxford.
Gyldenkerne, K. and Johansen, K. Y. (1970) Astr. Astrophys. Suppl.
    1, 129.
Hack, M. (1974) Astr. Astrophys. 36, 321.
Hack, M., Hutchings, J. B., Kondo, Y., McCluskey, G. E., Plavec, M.,
    and Polidan, R. S. (1974a) Nature 249, 534.
Hack, M., Hutchings, J. B., Kondo, Y., McCluskey, G. E., Plavec, M.,
    and Polidan, R. S. (1974b) Astrophys. J. 198, 453.
Hack, M., Hutchings, J. B. Kondo, Y., McCluskey, G. E., and Tulloch,
    M. K. (1976b) Astrophys. J. 206, 777.
Hack, M. and Job, F. (1965) Z. Astrophys. 62, 203.
Hack, M., van den Heuvel, E. P. J., Hoeckstra, R., de Jager, C.,
    and Sahade, J. (1976a) Astr. Astrophys. 50, 335.
Hackwell, J. A., Gehrz, R. D., and Smith, J. R. (1974) Astrophys.
    J. 192, 383.
Hall, D. S. (1967) Astr. J. 72, 301.
Hall, D. S. (1971) New Directions and New Frontiers in Variable
    Star Research (IAU Colloquium No. 15), Bamberg, p. 217.
Hall, D. S. (1972) Publ. astr. Soc. Pacific 84, 323.
Hall, D. S. (1975a) Acta astr. 25, 1.
Hall, D. S. (1975b) Acta astr. 25, 215.
Hall, D. S. (1975c) Acta astr. 25, 225.
Hall, D. S. (1976) Multiple Periodic Variables Stars, Ch. 15, ed.
    ed. W. S. Fitch, D. Reidel, Dordrecht. IAU Colloquium No. 29.
Hall, D. S. and Garrison (1972) Publ. astr. Soc. Pacific 84, 552.
Hall, D. S. and Keel, W. C. (1977) Acta astr. 27, 167.
Hall, D. S., Richardson, T. R., and Chambliss, C. R. (1976) Astr.
    J. 81, 138.
Hall, D. S. and Walter, K. (1974) Astr. Astrophys. 37, 263.
Hall, J. S. (1939) Astrophys. J. 90, 449.
Handbury, M. J. and Williams, I. P. (1976) Astrophys. Space Sci.
    45, 439.
Hansen, H. K. and McNamara, D. H. (1959) Astrophys. J. 130, 791.
Hardie, R. H. (1950) Astrophys. J. 112, 542.
Harnden, Jr., F. R., Fabricant, D., Topka, K., Flannery, B. P.,
    Tucker, W. H., and Gorenstein, P. (1977) Astrophys. J. 214, 418.
Harper, W. E. (1924) Publ. Dom. Astrophys. Obs. Victoria 3, 151.
Harper, W. E. (1933) J. R. astr. Soc. Can. 27, 146.

Hayakawa, S. and Matsuoko, M. (1964) Progr. Theoretical Phys. Suppl.
    No. 30, p. 204.
Hazelhurst, J. (1970) Mon. Not. R. astr. Soc. 149, 129.
Heise, J., Brinkman, A. C., Schrijver, D., Mewe, R., Groneschild,
    E., and den Boggende, A. (1975), quoted in Gorenstein and Tucker
    (1976).
Hellerich, J. (1922) Astr. Nachr. 216, 279.
Henize, J., Wray, J. D., Parsons, S. B., and Benedict, G. G. (1975)
    XIXth COSPAR meeting Varna, Bulgaria.
Herbst, W., Hesser, J. E., and Ostriker, J. P. (1974) Astrophys. J.
    193, 679.
Herczeg, T. (1956) Mitt. Budapest 41, 323.
Herczeg, T. (1960) Bonn Veröff No. 54.
Herczeg, T. (1975) Proc. Southwest Regional Conference for Astr.
    Astrophys. 1, 79.
Herczeg, T. and Frieboes-Conde, H. (1974) Astr. Astrophys. 30, 259.
Herczeg, T. and Schmidt, J. (1963) Z. Astrophys. 57, 254.
Herrero, V., Hjellming, R. M., and Wade, C. M. (1971) Astrophys. J.
    (Letters) 166, L19.
van den Heuvel, E. P. J. (1976) Structure and Evolution of Close
    Binary Systems, p. 35, eds. P. Eggleton et al., D. Reidel,
    Dordrecht, IAU Symp. No. 73.
Hill, G., Barnes, J. V., Hutchings, J. B., and Pearce, J. A. (1971)
    Astrophys. J. 168, 443.
Hill, G., Hilditch, R. W., Younger, F., and Fisher, W. A. (1975)
    Mem. R. astr. Soc. 79, 131.
Hiltner, W. A. (1947) Astrophys. J. 105, 214.
Hiltner, W. A. (1949) Astrophys. J. 110, 95.
Hiltner, W. A. (1950) Astrophys. J. 112, 477.
Hiltner, W. A. and Mook, D. E. (1966) Astrophys. J. 143, 1008.
Himple, K. (1936) Astr. Nachr. 261, 233.
Hjellming, R. M. (1976a) X-Ray Binaries (NASA SP-389) p. 233, eds.
    E. Boldt and Y. Kondo.
Hjellming, R. M. (1976b) X-Ray Binaries (NASA SP-389) p. 495, eds.
    E. Boldt and Y. Kondo.
Hjellming, R. M. (1976c) The Physics of Non-Thermal Radio Sources,
    p. 203, ed. G. Setti, D. Reidel, Dordrecht.
Hjellming, R. M. and Blankenship, L. (1973) Nature 243, 81; (see
    also IAU Inf. Bull. No. 2502).
Hjellming, R. M., Blankenship, L. C., and Balick, B. (1973a) Nature
    Phys. Sci. 242, 84.
Hjellming, R. M., Brown, R. L., and Blankenship, L. C. (1974)
    Astrophys. J (Letters) 194, L13.
Hjellming, R. M. and Hiltner, W. A. (1963) Astrophys. J. 137, 1080.
Hjellming, R. M. and Wade, C. M. (1970) Astrophys. J. (Letters)
    162, 1.
Hjellming, R. M. and Wade, C. M. (1971) Astrophys. J. (Letters)
    168, L21.
Hjellming, R. M. and Wade, C. M. (1971) Astrophys. J. (Letters)
    168, L115.
Hjellming, R. M. and Wade, C. M. (1973) Nature 242, 250.
Hjellming, R. M., Wade, C. M., Hughes, V. A., and Woodsworth, A.
    Nature 234, 138.
Hjellming, R. M., Webster, E., and Balick, B. (1972) Astrophys. J.
    (Letters) 178, 139.
Hogg, F. S. (1934) Publ. American astr. Soc. 8, 14.
Horák, T. (1968) Bull. astr. Inst. Csl. 19, 241.

168   Bibliography

Hosokawa, Y. (1955) Sendai astr. Rap. No. 42.
Hosokawa, Y. (1957) Sendai astr. Rap. No. 56.
Hosokawa, Y. (1959) Sendai astr. Rap. No. 70.
Hosokawa, Y. (1968) Sendai astr. Rap. No. 101.
Huang, S-S (1958) Publ. astr. Soc. Pacific 70, 473.
Huang, S-S (1963a) Astrophys. J. 138, 342.
Huang, S-S (1963b) Astrophys. J. 138, 471.
Huang, S-S (1965) Astrophys. J. 141, 976.
Huang, S-S (1973) Astrophys. Space Sci. 21, 263.
Huang, S-S and Brown, D. A. (1976) Astrophys. J. 208, 780.
Huang, S-S and Struve, O (1956) Astr. J. 61, 300.
Hughes, V. A. and Woodsworth, A. (1973) Nature Phys. Sci. 242, 116.
Humphreys, R. M. (1976) Astrophys. J. 206, 122.
Hutchings, J. B. (1973) Extended Atmospheres and Circumstellar
    Matter in Spectroscopic Binary Systems, p. 65, ed. A. H. Batten,
    D. Reidel, Dordrecht.
Hutchings, J. B., Cowley, A. P., and Redman, R. O. (1975) Astrophys.
    J. 201, 404.
Hutchings, J. B., Crampton, D., Glaspey, O., and Walker, G. A. H.
    (1973) Astrophys. J. 182, 549.
Iben, I. (1965a) Astrophys. J. 141, 993.
Iben, I. (1965b) Astrophys. J. 142, 1447.
Iben, I. (1966a) Astrophys. J. 143, 483.
Iben, I. (1966b) Astrophys. J. 143, 505.
Iben, I. (1967a) Astrophys. J. 147, 624.
Iben, I. (1967b) Astrophys. J. 147, 650.
Irwin. J. B. (1947) Astrophys. J. 106, 380.
Jachia, L. (1929a) Gaz. astr. 16, 19.
Jachia, L. (1928b) Gaz. astr. 16, 48.
Jameson, R. F. and Longmore, A. J. (1976) Mon. Not. R. astr. Soc.
    174, 217.
Johansen, K. T. (1969) Astr. Astrophys. 3, 179.
Johansen, K. T., Rudkjøbing, J., and Gyldenkerne, K. (1970) Astr.
    Astrophys. Suppl. 1, 149.
Johnson, H. M. (1977) preprint.
Jordan, F. C. (1916) Publ. Alegheny Obs. 3, 49.
Joss, P. C. and Rappaport, S. (1976) Nature 264, 219.
Joss, P. C. and Rappaport, S. (1977) Nature 265, 222.
Joy, A. H. (1930) Astrophys. J. 72, 41.
Joy, A. H. (1942) Publ. astr. Soc. Pacific 54, 35.
Joy, A. H. (1943) Publ. astr. Soc. Pacific 55, 283.
Joy, A. H. (1954a) Publ. astr. Soc. Pacific 66, 5.
Joy, A. H. (1954b) Astrophys. J. 120, 377.
Joy, A. H. (1956) Astrophys. J. 124, 317.
Jurkevich, I. (1970) Vistas in Astronomy, Vol. 12, ed. A. Beer,
    p. 63, Pergamon Press, Oxford.
Katz, J. I. (1975) Astrophys. J. 200, 298.
Keenan, P. C. and Wright, J. A. (1957) Publ. astr. Soc. Pacific
    69, 457.
Khosov, G. V. and Mineav, N. A. (1969) Trud. astr. Obs. Leningrad
    26, 55.
Kippenhahn, R., Kohl, K., and Weigert, A. (1967) Z. Astrophys.
    66, 58.
Kippenhahn, R. and Weigert, A. (1967) Z. Astrophys. 65, 251.
Kitamura, M. (1959) Publ. astr. Soc. Japan 11, 216.
Kitamura, M. (1960) Publ. astr. Soc. Japan 12, 1.
Kitamura, M. (1965) Adv. Astr. Astrophys. 3, 27.
Kitamura, M. (1967) Publ. astr. Soc. Japan 19, 194.

Kitamura, M. (1967) Publ. astr. Soc. Japan 19, 615.
Kitamura, M. (1974) Astrophys. Space Sci. 28, L17.
Kitamura, M. and Takahashi, C. (1962) Publ. astr. Soc. Japan 14, 44.
Kiyokawa, M. (1967) Publ. astr. Soc. Japan 19, 209.
Knipe, G. F. G. (1963) Republic Circ. 7, 21.
Koch, R. H. (1960) Astr. J. 65, 326.
Koch, R. H. (1968) Astr. J. 73, S102.
Koch, R. H. (1970) Vistas in Astronomy, Vol. 12, ed. A. Beer, p. 21
    Pergamon Press, Oxford.
Koch, R. H. (1972) Publ. astr. Soc. Pacific 84, 5.
Koch, R. H., Plavec, M., and Wood, F. B. (1970) Publ. Univ. Penn-
    sylvania astr. Ser. Vol. 11.
Kondo, Y. (1968) Highlights in Astronomy, ed. L. Perch. p. 432,
    D. Reidel, Dordrecht.
Kondo, Y. (1974) Astrophys. Space Sic. 27, 293.
Kondo, Y., Giuli, R. T., Modisette, J. L., and Rydgren, A. E. (1972)
    Astrophys. J. 176, 153.
Kondo, Y. and McCluskey, G. E. (1976) Structure and Evolution of
    Close Binary Systems, ed. P. Eggleton, et al., p. 277, D. Reidal,
    Dordrecht.
Kondo, Y., McCluskey, G. E., and Gulden, S. L. (1976) X-Ray Binaries
    (NASA SP-389) eds. E. Boldt and Y. Kondo, p. 499.
Kondo, Y., McCluskey, Jr., G. E. and Houck, T. E. (1971) New
    Directions and New Frontiers in Variable Star Research (IAU
    Colloquium No. 15), Bamberg, p. 308.
Kondo, Y., Morgan, T. H., and Modisette, J. L. (1975) Astrophys.
    J. (Letters) 196, L125.
Kondo, Y., Morgan, T. H., and Modisette, J. L. (1977) Publ. astr.
    Soc. Pacific (in press); Inf. Bull. Var. Stars. No. 1312.
Kondo, Y., Parsons, S. B., Henize, G. H., Wray, J. D., Benedict,
    G. F., and McCluskey, G. E. (1976) Astrophys. J. 208, 468.
Kopal, Z. (1938) Mon. Not. R. astr. Soc. 98, 448.
Kopal, Z. (1942) Proc. American Phil. Soc. 85, 399.
Kopal, Z. (1947) Harvard Circ. No. 450.
Kopal, Z. (1954) Joddrell Bank Ann. 1, 37
Kopal, Z. (1954) Observatory 74, 14.
Kopal, Z. (1955) les particules solides dans les astres, Mém. Soc.
    R. des. Sci. de Liège, 4éme séne, 15, 684.
Kopal, Z. (1955) Ann. Astrophys. 18, 379.
Kopal, Z. (1956) Ann. d'Astrophys. 19, 298.
Kopal, Z. (1957) Non-Stable Stars, ed. G. H. Herbig, p. 23, Cambridge
    Univ. Press.
Kopal, Z. (1959) Close Binary Systems, Chapman-Hall and John Wiley,
    New York.
Kopal, Z. (1965) Adv. Astr. Astrophys. 3, 89.
Kopal, Z. (1975) Astrophys. Space Sci. 34, 431.
Kopal, Z. (1976) Astrophys. Space Sci. 45, 269.
Kopal, Z. and Shapley, M. B. (1956) Jodrell Bank Ann. 1, 141.
Kraft, R. P. (1954) Astrophys. J. 120, 391.
Kraft, R. P. (1958) Astrophys. J. 127, 625.
Kraft, R. P. (1959) Astrophys. J. 130, 110
Kraft, R. P. (1962) Astrophys. J. 135, 408.
Kraft, R. P. (1963) Adv. Astr. Astrophys. 2, 43.
Kraft, R. P. (1964) Astrophys. J. 139, 457.
Kraft, R. P. (1964) Trans. int. astr. Un. 126, 519.
Kraft, R. P. (1967) Publ. astr. Soc. Pacific 79, 395.
Kraft, R. P., Krzeminski, W., and Mumford, G. S. (1969) Astrophys.
    J. 158, 589.

Kraft, R. P., Mathews, J., and Greenstein, J. L. (1962) Astrophys.
    J. 136, 321.
Krat, W. (1934) Astr. Zu. 11, 412.
Krat, W. (1936) Astr. Zu. 13, 527.
Krat, W. (1941) Poulkova Circ. No. 32, 87.
Kříž, S. (1974) Bull. astr. Inst. Czech. 25, 6.
Kron, G. E. (1947) Publ. astr. Soc. Pacific 59, 261.
Kron, G. E. and Gordon, K. C. (1943) Astrophys. J. 97, 311.
Kron, G. E. and Kron, K. C. (1950) Astrophys. J. 111, 454.
Kruszewski, A. (1964) Acta astr. 14, 231.
Kruszewski, A. (1964) Acta astr. 14, 241.
Krzemiński, W. (1965) Astrophys. J. 142, 1051.
Krzemiński, W. and Smak, J. (1971) Acta astr. 21, 133.
Kuhi, L. V. (1968) Wolf-Rayet Stars (NBS SP-307), ed. K. B. Gebbie
    and R. N. Thomas, p. 101.
Kuhi, L. V. (1973) Wolf-Rayet and High-Temperature Stars, ed. M.
    K. V. Bappu and J. Sahade, p. 205, D. Reidel, Dordrecht.
Kuhi, L. V. and Schweizer, F. (1970) Astrophys. J. (Letters) 160,
    L185.
Kuiper, G. P. (1941) Astrophys. J. 93, 133.
Kuiper, G. P. and Johnson, J. R. (1956) Astrophys. J. 123, 90.
Kuiper, G. P., Struve, O., and Strömgren, B. (1937) Astrophys. J.
    86, 570.
Kushwaha, R. S. (1957) Astrophys. J. 125, 242.
Kwee, K. K. (1958) Bull. astr. Inst. Netherl. 14, 131.
Labeyrie, A., Bonneau, D., Stachnik, R. V., and Gezari, D. Y.
    (1974) Astrophys. J. (Letters) 194, L147.
Larsson-Leander, G. (1953) Stockholm Obs. Ann. 17, No. 5.
Larsson-Leander, G. (1957) Stockholm Obs. Ann. 19, No. 8.
Larsson-Leander, G. (1957) Ark. Astr. 2, 135.
Larsson-Leander, G. (1959) Ark. Astr. 2, 301.
Larsson-Leander, G. (1961) Ark. Astr. 3, 25.
Larsson-Leander, G. (1969) Ark. Astr. 5, 253.
Lause, F. (1938) Astr. Nachr. 265, 205.
Lauterborn, D. (1969) Mass Loss from Stars, ed. M. Hack, p. 262,
    D. Reidel, Dordrecht.
Lauterborn, D. (1970) Astr. Astrophys. 7, 150.
Lauterborn, D. and Weigert, A. (1972) Astr. Astrophys. 18, 294.
Leavitt, H. (1903) Harvard Circ. No. 130.
Linnell, A. P. (1958) Astrophys. J. 127, 211.
Linnell, A. P. (1958) Astrophys. J. Suppl. 6, 109.
López, L. and Sahade, J. (1969) Bol. Assoc. Argentina Astr. No. 14,
    68.
Loreta, E. (1929) Gaz astr. 16, 10.
Loreta, E. (1930) Gaz astr. 17, 7.
Low, F. J. and Mitchell, R. I. (1964) Astrophys. J. 141, 327.
Ludendorff, H. (1903) Astr. Nachr. 164, 81.
Ludendorff, H. (1924) Sitz. Acad. Berlin, p. 49.
Lukatskaya, F. I. and Rubashevski, A. A. (1961) Var. Stars 13, 345.
Lucy, L. B. (1968a) Astrophys. J. 151, 1123.
Lucy, L. B. (1968b) Astrophys. J. 153, 877.
Lucy, L. B. (1973) Astrophys. Space Sci. 22, 381.
Lucy, L. B. (1976) Astrophys. J. 205, 208.
Lynden-Bell, D. and Pringle, J. E. (1974) Mon. Not. R. astr. Soc.
    168, 603.
Magalashvili, N. L. and Kumsishvili (1974) Abastumani Bull. No.
    45, 37.
Maggini (1918) Bull. astr. 35, 131.

Maraschi, L., Treves, A., and van den Heuvel, E. P. J. (1976) Nature 259, 292.

Markworth, N. L. (1977) Ph.D. Dissertation, Univ. of Florida, Gainesville, Florida, U.S.A.

Mason, K. O., Bechklin, E. E., Blankenship, L., Brown, R. L., Elias, J., Hjellming, R. M., Matthews, K., Murdin, P. G., Neugebauer, G., Sanford, P. W. and Willner, S. P. (1976) Astrophys. J. 207, 78.

Mason, K. O., Murdin, P. G., and Visvanathan, N. (1977) IAU Circ. No. 3054.

Mathis, J. S. (1967) Astrophys. J. 149, 619.

Mathis, J. S. and Odell, A. P. (1973) Astrophys. J. 180, 517.

Matilsky, T., Bradt, H. V., Buff, J., Clark, G. W., Jernigan, J. G., Joss, P. C., Lanfer, B., and McClintock, J. (1976) Astrophys. J. (Letters) 210, L127.

Matteson, J. L. Mushotzky, R. F., Paciesas, W. S., and Laros, J. G. (1976) X-Ray Binaries (NASA SP-389), ed. E. B. Boldt and Y. Kondo, p. 407.

Mauder, H. (1966) Kleine Veröff. Remeis Sternw. No. 38.

Maury, A. C. (1897) Harvard Ann. 28.

McCluskey, Jr., G. E., Kondo, Y., and Morton, D. S. (1975) Astrophys. J. 201, 607.

McCluskey, Jr., G. E. and Wood, F. B. (1970) Vistas in Astronomy, Vol. 12, ed. A. Beer, p. 257, Pergamon Press, Oxford.

McKellar, A. and Petrie, R. M. (1952) Mon. Not. R. astr. Soc. 112, 641.

McKellar, A. and Petrie, R. M. (1958) Publ. Dom. Astrophys. Obs., Victoria, 11, No. 1.

McKellar, A. Odgers, G. J., McLaughlin, D. B., and Aller, L. H. (1952) Nature 109, 990; Contr. Dom. Astrophys. Obs., Victoria, No. 24.

McKellar, A. and Wright, K. O. (1957) J. R. astr. Soc. Can. 51, 75.

McKellar, A., Wright, K. O., and Francis, J. D. (1957) Publ. astr. Soc. Pacific 69, 442.

McLaughlin, D. B. (1924) Astrophys. J. 60, 22.

McLaughlin, D. B. (1936) Harvard Announcement Card No. 397.

McLaughlin, D. B. (1948) Astrophys. J. 108, 237.

McLaughlin, D. B. (1950a) Publ. astr. Soc. Pacific 62, 13.

McLaughlin, D. B. (1950b) Astrophys. J. 111, 449.

McLaughlin, D. B. (1952a) Publ. astr. Soc. Pacific 64, 173.

McLaughlin, D. B. (1952b) Astrophys. J. 116, 546.

McLaughlin, D. B. (1954) Astr. J. 59, 347.

McLaughlin, D. B. (1960) Stellar Atmospheres, ed. J. L. Greenstein, p. 585, Univ. Chicago Press.

Mendéz, R. H., Munch, G., and Sahade, J. (1975) Publ. astr. Soc. Pacific 87, 305.

Menzel, D. M. (1936) Harvard Circ. No. 417.

Merrill, J. E. (1950, 1953a) Contr. Princeton Obs. No. 23.

Merrill, J. E. (1953b) Contr. Princeton Obs. No. 24.

Merrill, J. E. (1970) Vistas in Astronomy, Vol. 12, ed. A. Beer, p. 43, Pergamon Press, Oxford.

Merrill, P. W. (1958) Étoiles a raies d'émission, p. 436, Cointe-Sclessin.

Miczaika, G. R. (1953) Z. Astrophys. 33, 1.

Milne, E. A. (1926) Mon. Not. R. astr. Soc. 87, 43.

Mochanacki, S. W. and Doughty, N. A. (1972a) Mon. Not. R. astr. Soc. 156, 51.

Mochanacki, S. W. and Doughty, N. A. (1972b) Mon. Not. R. astr.
    Soc. 156, 243.
Morris, S. C. (1962) J. R. astr. Soc. Can. 56, 210.
Morton, D. C. (1960) Astrophys. J. 132, 146.
Morton, D. C., Jenkins, E. B., and Brooks, N. H. (1969) Astrophys.
    J. 155, 875.
Moss. D. L. and Whelan, J. A. J. (1970) Mon. Not. R. astr. Soc.
    149, 447.
Mullan, D. J. (1974) Astrophys. J. 192, 149.
Mumford, G. S. (1962) Sky and Telesc. 23, 135
Mumford, G. S. (1964) Astrophys. J. 139, 476.
Mumford, G. S. (1967) Publ. astr. Soc. Pacific 79, 282.
Münch, G. (1950) Astrophys. J. 112, 266.
Myers, G. W. (1898) Astrophys. J. 7, 1.
Napier, W. McD. (1968) Astrophys. Space Sci. 2, 61.
Napier, W. McD. (1971) Astrophys. Space Sci. 11, 475.
Nariai, K. (1963) Publ. astr. Soc. Japan 15, 7.
Nather, R. E. and Warner, B. (1971) Mon. Not. R. astr. Soc. 152,
    209.
Nelson, B. and Duckworth, E. (1968) Publ. astr. Soc. Pacific 80,
    562.
Nelson, B. and Young, A. (1976) Structure and Evolution of Close
    Binary Systems, ed. P. Eggleton, et al., p. 141, D. Reidel,
    Dordrecht.
Nha, I. S. (1975) Astr. J. 80, 232.
Niemelä, V. S. (1973) Publ. astr. Soc. Pacific 85, 220.
Niemelä, V. S. (1976) Astrophys. Space Sci. 45, 191.
Niemelä, V. S. (1977) Private Communication.
Nowikov, I. D. and Zeldovich, Ya. B., (1966) Nuovo Cimento Suppl.
    4, 810, addendums 2.
Odell, A. P. (1974) Astrophys. J. 192, 417.
Okazaki, A. (1977) Publ. astr. Soc. Japan 29, 289.
Oliver, J. P. (1974) Ph.D. Dissertation, Univ. of California, Los
    Angeles, California, U.S.A.
Oliver, J. P. (1975) Publ. astr. Soc. Pacific 87, 695.
Olson, E. C. (1974) I.A.U. Circ. No. 2716.
Olson, E. C. (1976a) Astrophys. J. 204, 141.
Olson, E. C. (1976b) Astrophys. J. Suppl. Ser. 31, 1.
Olson, E. C. (1978) Astrophys. J. (in press).
Osaki, Y. (1965) Publ. astr. Soc. Japan 17, 79.
Ostriker, J. P., Molray, R., Weaver, R., and Yahil, A. (1976)
    Astrophys. J. 208, 61.
Ovenden, M. W. (1956) Vistas in Astronomy, Vol. 2, ed. A. Beer,
    p. 1193, Pergamon Press, Oxford.
Ovenden, M. W. (1956) Vistas in Astronomy, Vol. 2, ed. A. Beer,
    p. 1195, Pergamon Press, Oxford,.
Ovenden, M. W. (1970) Vistas in Astronomy, Vol. 12, ed. A. Beer,
    p. 135, Pergamon Press, Oxford.
Owen, R. N., Jones, T. W., and Gibson, D. M. (1976) Astrophys. J.
    (Letters) 210, L27.
Paczyński, B. (1965) Acta astr. 15, 89.
Paczyński, B. (1967a) Acta astr. 17, 355.
Paczyński, B. (1967b) On the Evolution of Double Stars (Obs. R.
    de Belgique Comm. Ser. B, No. 17), ed. J. Dommanget, p. 111.
Paczyński, B. (1970a) Mass Loss and Evolution in Close Binaries,
    ed. K. Gyldenkerne and R. M. West, p. 139, Copenhagen Univ.
    Publ. Fund.
Paczyński, B. (1970b) Acta astr. 20, 195.

Paczyński, B. (1971) Ann. Review Astr. Astrophys. 9, 183.
Paczyński, B. and Ziólkowski, J. (1967) Acta astr. 17, 7.
Paczyński, B. Ziólkowski, J., and Zithow, A. (1969) Mass Loss from Stars, ed. M. Hack, p. 237, D. Reidel, Dordrecht.
Parsons, S. B., Wray, J. D., Kondo, Y., Henize, K. G., and Benedict, G. F. (1976) Astrophys. J. 203, 435.
Payne-Gaposchkin, C. (1957) The Galactic Novae, North-Holland, Amsterdam.
Payne-Gaposchkin, C. (1977) Novae and Related Stars, ed. M. Friejung, p. 3, D. Reidel, Dordrecht.
Perek, L. (1968) Highlights of Astronomy, p. 393, D. Reidel, Dordrecht.
Peters, G. J. (1973) Extended Atmospheres and Circumstellar Matter in Spectroscopic Binary Systems, ed. A. H. Batten, p. 174, D. Reidel, Dordrecht.
Petty, A. F. (1973) Astrophys. Space Sci. 21, 189.
Pfeiffer, R. J. and Koch, R. H. (1973) IAU Inf. Bull. Var. Stars No. 780.
Pfeiffer, R. J. and Koch, R. H. (1977) Publ. astr. Soc. Pacific 89, 147.
Pfeiffer, R. J. and Koch, R. H. (1977) Astr. J. 89, 147.
Pickering, E. C. (1880) Proc. Amer. Acad. Arts Sci. 16, 257.
Pickering, E. C. (1896) Harvard Circ. No. 11.
Pickering, E. C. (1904) Ann. Harv. Coll. Obs. 46, 121.
Pickering, E. C. (1904) Ann. Harv. Coll. Obs. 46, 172.
Piirola, V. (1975) IAU Inf. Bull. Var. Stars No. 1061.
Pillans. H. (1956) Publ. astr. Soc. Pacific 67, 340.
Piotrowski, S. L. (1937) Acta astr. a 4, 1.
Piotrowski, S. L. (1964) Acta astr. 14, 251.
Piotrowski, S. L. and Rozyczka, M. (1973) Var. Stars 19, 107.
Piotrowski, S. L, Rucinski, S. M., and Semeniuk, I. (1974) Acta astr. 24, 387.
Plaut, L. (1953) Publ. Kapteyn astr. Lab. No. 55.
Plavec, M. (1958) Étoiles á raies d émission, p. 411, Institut d'Astrophysique, Cointe-Scelessin, p. 411.
Plavec, M. (1964) Bull. astr. Inst. Csl. 15, 165.
Plavec, M. (1966) Trans. IAU 12B, 508.
Plavec, M. (1967) On the Evolution of Double Stars, ed. J. Dommanget, p. 83, Commun. Obs. r. Belgique, Ser B. No. 17.
Plavec, M. (1968) Advances in Astr. Astrophys. 6, 201.
Plavec, M. (1970) Publ astr. Soc. Pacific 82, 957.
Plavec, M. (1973) Extended Atmospheres and Circumstellar Matter in Spectroscopic Binary Systems, ed. A. H. Batten, p. 216, D. Reidel, Dordrecht.
Plavec, M. and Kriz, S. (1965) Bull. astr. Inst. Csl. 16, 297.
Plavec, M. and Polidan, R. S. (1975) Nature 253, 173.
Plavec, M. and Polidan, R. S. (1976) Structure and Evolution of Close Binary Systems, ed. P. Eggleton, et al., p. 289, D. Reidel, Dordrecht.
Plavec, M. Senhal, L., and Mikulas, J. (1964) J. Bull. astr. Inst. Csl. 15, 171.
Plavec, M., Ulrich, K. R., and Polidan, R. S. (1973) Publ. astr. Soc. Pacific 85, 769.
Pollard, H. (1966) Mathematical Introduction to Celestial Mechanics, Printice-Hall, Englewood Cliffs.
Popper, D. M. (1970) Spectroscopic Astrophysics, ed. G. H. Herbig, p. 155, Univ. California Press.

Popper, D. M. (1970) Mass Loss and Evolution in Close Binaries,
    ed. K. Gyldenkerne and R. M. West, p. 104, Copenhagen Univ.
    Publ. Fund.
Popper, D. M. and Plavec, M. (1974) Bull. Am. astr. Soc. 6, 334.
Popper, D. M. and Plavec, M. (1976) Astrophys. J. 205, 642.
Popper, D. M. and Ulrich, R. K. (1976) Astrophys. J. (Letters)
    212, L131.
Prendergast, K. (1960) Astrophys. J. 132, 162.
Prendergast, K. and Burbidge, G., (1968) Astrophys. J. (Letters)
    15, L83.
Prendergast, K. and Taam, R. (1974) Astrophys. J. 189, 125.
Pringle, J. E. (1977) Mon. Not. R. astr. Soc. 178, 195.
Proctor, D. D. and Linnell, A. P. (1972) Astrophys. J. Suppl.
    Ser 24, 449.
Purton, C. R., Feldman, P. A., and Marsh, K. A., (1973) Nature
    Phys. Sci. 245, 5.
Pustylnik, I. B. (1967) Astrofiz 3, 69.
Pustylnik, I. and Toomasson, L. (1973) Astrophys. Space Sci. 21,
    495.
Rafert, J. B. (1977) Ph.D. Dissertation, University of Florida,
    Gainesville, Florida, U.S.A.
Rappaport, S., Cash, W., Doxley, R., McClintock, J., and Moore, J.,
    (1976) Astrophys. J. (Letters) 187, L5.
Rappaport, S. and Joss, P. C. (1977a) Nature 266, 123.
Rappaport, S. and Joss, P. C. (1977b) Nature 266, 683.
Refsdal, S. and Weigert, A. (1969) Astr. Astrophys. 1, 167.
Refsdal, S. and Weigert, A. (1971) Astr. Astrophys. 13, 367.
Rhombs, C. E. and Fix, J. D. (1976) Astrophys. J. 209, 821.
Ritter, H. (1976) Mon. Not. R. astr. Soc. 175, 279.
Roach, F. E. and Wood, F. B. (1952) Ann. Astrophys. 15, 21.
Roberts, A. W. (1903) Mon. Not. R. astr. Soc. 63, 528.
Roberts, A. W. (1908) Mon. Not. R. astr. Soc. 68, 490.
Robinson, E. L. (1973a) Astrophys. J. 180, 121.
Robinson, E. L. (1973b) Astrophys. J. 186, 347.
Robinson, E. L. (1976) Astrophys. J. 203, 485.
Robinson, E. L. (1976) Ann. Reviews Astr. Astrophys. 14, 119.
Rossiter, R. A. (1933) Publ. Michigan Obs. 5, 67.
Rudkjobing (1959) Ann. Astrophys. 22, 111.
Rucinski, S. M. (1969a) Acta astr. 19, 125.
Rucinski, S. M. (1969b) Acta astr. 19, 245.
Rucinski, S. M. (1970a) Acta astr. 20, 249.
Rucinski, S. M. (1970b) Acta astr. 20, 327.
Rucinski, S. M. (1971) Acta astr. 21, 455.
Rucinski, S. M. (1973a) Acta astr. 23, 79.
Rucinski, S. M. (1973b) Acta astr. 23, 301.
Rucinski, S. M. (1974) Acta astr. 24, 119.
Russell, H. N. (1912a) Astrophys. J. 35, 315.
Russell, H. N. (1912b) Astrophys. J. 36, 54.
Russell, H. N. (1912c) Astrophys. J. 36, 133.
Russell, H. N. (1928) Mon. Not. R. astr. Soc. 88, 641.
Russell, H. N. (1942) Astrophys. J. 95, 345.
Russell, H. N. (1945) Astrophys. J. 102, 1.
Russell, H. N. (1946) Astrophys. J. 104, 153.
Russell, H. N. (1948a) Astrophys. J. 108, 53; 388.
Russell, H. N. (1948b) Harvard Circ. No. 452.
Russell, H. N. and Merrill, J. E. (1952) Contr. Princeton Univ. Obs.
    No. 26.
Russell, H. N. and Shapley, H. (1912a) Astrophys. J. 36, 239.

Russell, H. N. and Shapley, H. (1912b) Astrophys. J. 36, 385.
Sahade, J. (1945) Astrophys. J. 102, 474.
Sahade, J. (1949) Astrophys. J. 109, 439.
Sahade, J. (1952) Astrophys. J. 116, 35.
Sahade, J. (1955) Publ. astr. Soc. Pacific 67, 348.
Sahade, J. (1958) Astr. J. 63, 52.
Sahade, J. (1958) Publ. astr. Soc. Pacific 70, 319.
Sahade, J. (1958a) Etoiles á raies d'emission, Institut d'Astro-
    physique, Cointe-Scelessin, p. 405.
Sahade, J. (1958b) Etoiles á raies d'emission, Institut d'Astro-
    physique, Cointe-Scelessin, p. 46.
Sahade, J. (1959) Publ. astr. Soc. Pacific 71, 151.
Sahade, J. (1959) Liege Symposium, Modèles d'étoiles et l'évolution
    stellaire, p. 76.
Sahade, J. (1960a) Stellar Atmospheres (Stars and Stellar Systems
    VI), ed. J. L. Greenstein, p. 466, University of Chicago
    Press.
Sahade, J. (1960b) Modéles d'étoiles et évolution stellaire,
    Universite de Liège, p. 76.
Sahade, J. (1962) Symposium on Stellar Evolution, ed. J. Sahade,
    Observatorio Astronomico, LaPlata, p. 185.
Sahade, J. (1963) Ann. Astrophys. 26, 80.
Sahade, J. (1965) Kleine Veröff, Remeis-Sternw, 4, 140.
Sahade, J. (1965) Observatory 85, 214.
Sahade, J. (1966) Trans. IAU 12B, 494.
Sahade, J. (1966) Modern Astrophysics, ed. M. Hack, p. 219,
    Gauthier-Villars, Paris.
Sahade, J. (1968) Wolf-Rayet Stars (NBS SP-307), ed. K. B. Gebbie
    and R. N. Thomas, p. 89.
Sahade, J. (1969) Mass Loss from Stars, ed. M. Hack, p. 156, D.
    Reidel, Dordrecht.
Sahade, J. (1973) Extended Atmospheres and Circumstellar Matter in
    Spectroscopic Binary Systems, ed. A. H. Batten, p. 286., D
    Reidel, Dordrecht.
Sahade, J. (1975) Astrophysics, ed. C. de Loore and L. Howziaux,
    Vrije Universiteit Brussel.
Sahade, J. (1976a) Structure and Evolution of Close Binary Systems,
    ed. R. Eggleton et al., p. 1, D. Reidel, Dorhrecht.
Sahade, J. and Albano J. (1970) Astrophys. J. 162, 905.
Sahade, J. and Cesco, C. U. (1945) Astrophys. J. 101, 235.
Sahade, J. and Frieboes-Conde, H. (1965) Astrophys. J. 141, 652.
Sahade, J., Huang, S-S, Struve, O., and Zebergs, V. (1959) Trans.
    American Phil. Soc. 49, 1.
Sahade, J. and Ringuelet, A. (1970) Vistas in Astronomy, Vol. 12,
    ed. A. Beer, p. 143, Pergamon Press, Oxford.
Sahade, J. and Struve, O. (1945) Astrophys. J. 102, 480.
Sahade, J. and Struve, O. (1957) Astrophys. J. 126, 87.
Sahade, J. and Wallerstein, G. (1958) Publ. astr. Soc. Pacific 70,
    207.
Saito, M. (1965) Publ. astr. Soc. Japan 17, 107.
Saito, M. (1970) Publ. astr. Soc. Japan 22, 455.
Saito, M. (1973) Astrophys. Space Sci. 22, 133.
Saito, M. and Kawabata, S. (1976) Astrophys. Space Sci. 45, 63.
Saito, M., Sato, H., and Sato, N. (1972) Tokyo astr. Bull. No. 219.
Sandage, A. R., Osmer, P., Giasconi, R., Gorenstein, P., Bursky,
    H., Waters, J., Bradt, H., Garmire, G., Brunkantan, B. V., Oda,
    M., Osawa, K., and Jugaku, J. (1966) Astrophys. J. 146, 316.

Sanford, R. F. (1949) Astrophys. J. 109, 81.
Sanyal, A. (1976) Astrophys. J. 210, 853.
Sato, H. (1971) Publ. astr. Soc. Japan 23, 335.
Sato, H. and Saito, M. (1969) Tokyo astr. Bull. No. 192.
Sato, H. and Saito, M. (1973) Tokyo astr. Bull. No. 227.
Sawyer, A. J. (1887) Astr. J. 7, 119.
Schaube, S. (1924) Bull. Poulkov 10, 31.
Schneller, H. (1928) Astr. Nachr. 233, 361.
Schneller, H. and Plaut, L. (1932) Astr. Nachr. 245, 387.
Schnopper, H. W., Delvaille, J. P., Epstein, A., Helmken, H.,
    Murray, S. S., Clark, G., Jernigan, G., and Donsey, R. (1976)
    Astrophys. J. (Letters) 210, L75.
Schönberg, E. and Jung, B. (1938) Astr. Nachr. 265, 221.
Schuerman, D. W. (1972) Astrophys. Space Sci. 19, 351.
Schwarzschild, M. (1956) Structure and Evolution of the Stars,
    p. 198, Princeton Univ. Press.
Seaquist, E. R. (1976a) Astrophys. J. (Letters) 203, L35.
Seaquist, E. R. (1976b) Private Communication.
Seaquist, E. R. and Gregory, P. C. (1973) IAU Circ. No. 2563.
Seaquist, E. R., Gregory, P. C., Perley, R. A., Becker, R. H.,
    Carlson, J. B., Kundu, M. R., Bignell, R. C., and Dickell,
    J. R. (1974) Nature 251, 396.
Seggewiss, W. (1974) Publ. astr. Soc. Pacific 86, 670.
Semeniuk, I. (1968) Acta astr. 18, 1.
Seminiuk, I. and Paczyński, B. (1968) Acta astr. 18, 33.
Sen, H. K. (1948) Proc. Nat. Acad. Sci. 34, 311.
Serkowski, K. (1971) New Direction and New Frontiers in Variable
    Stars Research (IAU Colloquium No. 15), Bamberg, p. 11.
Shakhovskoj, N. M. (1962) Astr. Zh. 39, 755 [Trans. Soviet Astr.
    6, 587].
Shakhovskoj, N. M. (1964) Astr. Zh. 41, 1042 [Trans. Soviet Astr.
    8, 833].
Shapley, H. (1915) Contr. Princeton Univ. Obs. No. 3.
Shapley, H. (1922) Bull. Harvard Obs. No. 767.
Shklovsky, I. S. (1967) Astrophys. J. (Letters) 148, L1.
Shu, H., Lubos, S. H., and Anderson, L. (1976) Astrophys. J. 209,
    536.
Shulman, S., Friedman, H., Fritz, G., Henry, R. C., and Yentis,
    D. J. (1975) Astrophys. J. (Letters) 199, L101.
Shulov, O. S. (1966) Astr. Circ. U. S. S. R. No. 385.
Shulov, O. S. (1967) Trudy Leningrad Astr. Obs. 24, 38.
Sistero, R. F. (1968) Publ. astr. Soc. Pacific 80, 474.
Sistero, R. F. (1971) Bull. astr. Inst. Csl. 22, 188.
Sitterly, B. W. (1928) Pop. Astr. 36, 297.
Sitterly, B. W. (1930) Contr. Princeton Univ. Obs. No. 11.
Sletteback, A. (1976) Be and Shell Stars, D. Reidel, Dordrecht.
Smak, J. (1961) Acta astr. 11, 171.
Smak, J. (1962) Acta astr. 12, 28.
Smak, J. (1970) Acta astr. 20, 311.
Smak, J. (1971) Acta astr. 21, 15.
Smak, J. (1971) New Directions and New Frontiers in Variable Star
    Research, (IAU Colloquium No. 15) Bamberg, p. 248.
Smak, J. (1972) Acta astr. 22, 1.
Smith, H. E., Margon, B., and Conti, P. S. (1973) Astrophys. J.
    182, 549.
Smith, L. F. (1972) Scientific Results from the OAO-2 (NASA SP-310)
    ed. A. D. Code, p. 429.
Sobieski, S. (1965a) Astrophys. J. Suppl. Ser. 12, 263.

Sobieski, S. (1965b) Astrophys. J. Suppl. Ser. 12, 276.
Sparks, W. M. and Stecher, T. P. (1974) Astrophys. J. 188, 149.
Starrfield, S., Sparks, W. M., and Truran, J. W. (1974) Astrophys.
    J. Suppl. 28, 247.
Starrfield, S., Sparks, W. M. and Truran, J. W. (1976) Structure
    and Evolution of Close Binary Systems, ed. P. Eggleton et al.
    p. 155, D. Reidel, Dordrecht.
Stebbins. J. (1910) Astrophys. J. 32, 185.
Stebbins, J. (1911) Astrophys. J. 33, 395.
Stebbins, J. (1921) Astrophys. J. 53, 105.
Stebbins, J. (1922) Pop Astr. 30, 233.
Stebbins, J. and Gordon, K. C. (1975) Astrophys. Space Sci. 33, 481.
Stecher, T. P. (1968) Wolf-Rayet Stars (NBS SP-307) ed. K. B. Gebbie
    and R. N. Thomas, p. 65.
Struve, O. (1934) Observatory 57, 265.
Struve, O. (1935) Pop Astr. 43, 493.
Struve, O. (1941) Astrophys. J. 93, 104.
Struve, O. (1944) Astrophys. J. 99, 22.
Struve, O. (1944a) Astrophys. J. 99, 22.
Struve, O. (1944b) Astrophys. J. 99, 89.
Struve, O. (1944c) Astrophys. J. 99, 295.
Struve, O. (1947) Astrophys. J. 106, 255.
Struve, O. (1948) Ann. Astrophys. 11, 117.
Struve, O. (1949) Mon. Not. R. astr. Soc. 109, 487.
Struve, O. (1950a) Stellar Evolution, Princeton Univ. Press.
Struve, O. (1950b) Stellar Evolution, p. 168, Princeton Univ. Press.
Struve, O. (1952) Publ. astr. Soc. Pacific 64, 20.
Struve, O. (1953) Publ. astr. Soc. Pacific 65, 84.
Struve, O. (1955) Publ. astr. Soc. Pacific 67, 348.
Struve, O. (1955) Sky Telesc. 14, 275.
Struve, O. (1955) Les particules solides dans les astres., Mem.
    Soc. R. Sci. de Lieg, 4éme. série, 15, 686.
Struve, O. (1956) Publ. astr. Soc. Pacific 68, 27.
Struve, O. (1957) Occasional Notes R. astr. Soc. 3, 161.
Struve, O. (1957) Sky Telesc. 17, 70.
Struve, O. (1957) Non-Stable Stars, ed. G. H. Herbig, p. 93,
    Cambridge Press.
Struve, O. (1958) Publ. astr. Soc. Pacific 69, 41.
Struve, O. (1962) Symposium on Stellar Evolution, ed. J. Sahade,
    p. 225, University of La Plata Obs.
Struve, O. and Huang, S-S, (1958) Astrophys. J. 127, 148.
Struve, O. and Pillans, H., (1957) Publ. astr. Soc. Pacific 69, 27.
Struve, O. Pillans, H., and Zebergs, V. (1958) Astrophys. J. 128,
    287.
Struve, O. and Sahade, J. (1957a) Publ. astr. Soc. Pacific 69, 41.
Struve, O. and Sahade, J. (1957b) Astrophys. J. 125, 689.
Struve, O. and Sahade, J. (1958) Publ. astr. Soc. Pacific 70, 111.
Struve, O. and Sahade, J. (1958) Publ. astr. Soc. Pacific 70, 313.
Struve, O., Sahade, J., and Huang, S-S. (1957) Publ. astr. Soc.
    Pacific 69, 342.
Struve, O. and Smith, B. (1950) Astrophys. J. 111, 27.
Struve, O. and Wade, M. J. (1960) Publ. astr. Soc. Pacific 72, 403.
Struve, O. and Zebergs, V. (1961) Astrophys. J. 134, 161.
Struve, O. and Zebergs, V. (1962) Astronomy of the 20th Century.
    p. 305, Macmillan, New York.
Stothers, R. (1974) Astrophys. J. 194, 651.
Stub, H. (1972) Astr. Astrophys. 20, 161.

Swings, P. (1970) Spectroscopic Astrophysics, ed. G. H. Herbig,
    p. 189, Univ. of California Press.
Szebehely, Y. (1967a) Theory of Orbits, p. 297, Academic Press.
Szebehely, V. (1967b) Theory of Orbits, p. 587, Academic Press.
Takeda, S. (1934) Kyoto Mem. Ser. A. 17, 197.
Tananbaum, H., Gursky, H., Kellogg, E., Giacconi, R., Jones, C.
    (1972) Astrophys. J. (Letters) 177 , L5.
Taylor, J. H. and Hulse, R. A. (1974) IAU Circ. No. 2704.
Thackeray, A. D. (1973) Extended Atmospheres and Circumstellar
    Matter in Spectroscopic Binary Systems. ed. A. H. Batten, p. 267,
    D. Reidel, Dordrecht.
Thomas, H. C. (1976) Max Plank Institut fur Physik und Astrophysik
    MPl-PAE/Astro No. 100.
Trimble, V., Rose, W. K., and Webber, J. (1973) Mon. Not. R. astr.
    Soc. 162, 1P.
Tschudovitchev, N. (1950) Astr. Circ. Kazan, Nr. 100, 14.
Tsesevich, V. P. (1973) Eclipsing Variable Stars, Halsted Press
    New York.
Tsumeni, H., Matsuoka, M., and Takagishi, K. (1977) Astrophys. J.
    (Letters) 211, L15.
Ulrich, R. K. and Burger, H. L. (1976) Astrophys. J. 206, 509.
Underhill, A. B. (1954) Mon. Not. R. astr. Soc. 114, 558.
Underhill, A. B. (1969a) Mass Loss from Stars, ed. M. Hack, p. 17,
    D. Reidel, Dordrecht.
Underhill, A. B. (1969b) Mass Loss from Stars, ed. M. Hack, p. 229,
    D. Reidel, Dordrecht.
Ureche, V. (1972) Studii Cerc. Astr. 17, 213.
Vanbeveren, D. (1977) Astr. Astrophys. 54, 877.
van den Heuvel, E. P. J., (1973) Nature Phys. Sci. 242, 71.
van den Heuvel, E. P. J., (1976) Structure and Evolution of Close
    Binary Systems, ed. P. Eggleton et al., p. 35, D. Reidel,
    Dordrecht.
van der Hucht, K. A. (1975) Phil. Trans. R. Soc. London A 279, 451.
Van t'Veer, F. (1972) Astr. Astrophys. 19, 337.
Vasileva, A. A. (1952) Bjull. Stalinabad Astr. Obs. No. 4.
Vilhu, O. (1973) Astr. Astrophys. 26, 267.
Vinter-Hansen, J. M. (1944) Astrophys. J. 100, 8.
Wade, C. M. and Hjellming, R. M. (1972) Nature 235, 270.
Walker, M. F. (1954) Publ. astr. Soc. Pacific 66, 230.
Walker, M. F. and Herbig, G. H. (1954) Astrophys. J. 120, 278.
Wallerstein, G., Greene, T. F., and Tomley, L. J. (1967) Astrophys.
    J. 150, 245.
Walraven, J. H. (1969) Inf. Bull. Sth. Hemisphere No. 15, p. 40.
Walter, K. (1933) Veröff, Königsberg No. 3.
Walter, K. (1948) Astr. Nachr. 276, 225.
Warner, B. (1971) New Directions and New Frontiers in Variable
    Star Research (IAU Colloquium No. 15), Bamberg, p. 267.
Warner, B. (1974a) Mon. Not. R. astr. Soc. 167, 61P.
Warner, B. (1974b) Mon. Not. R. astr. Soc. 168, 235.
Warner, B. (1976) Structure and Evolution of Close Binary Systems.
    ed. P. Eggleton et al. p. 85, D. Reidel, Dordrecht.
Warner, B. and Nather, R. E. (1971) Mon. Not. R. astr. Soc. 152, 219.
Warner, B. and Nather, R. E. (1972) Sky Telesc. 43, 82.
Warner, B. and Robinson, E. L. (1972) Nature 239, 2.
Weigert, A. (1968) Highlights of Astronomy, ed. L. Perek, p. 414,
    D. Reidel, Dordrecht.
Weiler, E. J. (1975) IAU Inf. Bull. Var. Stars, No. 1014.

Wellmann, P. (1953) Nachr. Astr. Zetralstelle 7, 7.
Wendell, O. C. (1907) Astr. Nachr. 175, 357.
Wendell, O. C. (1909) Ann. Harv. Coll. Obs. 69, 66.
Whelan, J. A. J., Worden, S. P., and Mochnacki, S. W. (1973)
    Astrophys. J. 183, 133.
Wilson, O. C. (1939) Publ. astr. Soc. Pacific 51, 55.
Wilson, O. C. (1940) Astrophys. J. 91, 379.
Wilson, O. C. (1960) Stars and Stellar Systems, Vol. VI,
    Stellar Atmospheres, ed. J. L. Greenstein, Ch. 11.   .
Wilson, O. C. and Abt, H. A. (1954) Astrophys. J. Suppl. 1, 1.
Wilson, R. E. (1971) Astrophys. J. 170, 529.
Wilson, R. E. (1974) Astrophys. J. 189, 319.
Wilson, R. E., DeLuccia, M. R., Johnston, K., and Mango, S. A.
    (1972) Astrophys. J. 177, 191.
Wilson, R. E. and Devinney, E. J. (1971) Astrophys. J. 166, 605.
Wilson, R. E. and Sabatino, S. (1975) Astrophys. J. 203, 182.
Wilson, R. E. and Starr, T. C. I. (1976) Mon. Not. R. astr. Soc.
    176, 625.
Wolf, C. J. E. and Rayet, G. (1867) Compt. Rend. 65, 292.
Wood, D. B. (1962) Trans. IAU 11B, 386.
Wood, D. B. (1971) Astr. J. 76, 701.
Wood, D. B. (1976) Astr. J. 81, 855.
Wood, F. B. (1940) Astr. J. 48, 185.
Wood, F. B. (1946) Contr. Princeton Univ. Obs. No. 21.
Wood, F. B. (1950) Astrophys. J. 112, 196.
Wood, F. B. (1953a) Astronomical Photoelectric Photometry, ed.
    F. B. Wood, Amer. Assoc. Adv. Sci, Washington, D. C.
Wood, F. B. (1953b) Astr. J. 58, 51.
Wood, F. B. (1957a) The Present and Future of the Telescope of
    Moderate Size, ed. F. B. Wood, Univ. of Pennsylvania Press,
    Philadelphia.
Wood, F. B. (1957b) Non-Stable Stars, IAU Symp. No. 3, ed. G. H.
    Herbig, p. 144, Cambridge Univ. Press.
Wood, F. B. (1964) Vistas in Astronomy, Vol. 5, ed. A. Beer,
    p. 119, Pergamon Press, Oxford.
Wood, F. B. and Austin, R. R. D. (1978) in press.
Wood, F. B. and Lewis, E. M. (1954) Astr. J. 59, 119.
Wood, F. B. and Richardson, R. R. (1964) Astr. J. 69, 297.
Woodsworth, A. W. and Hughes, V. A. (1976) Mon. Not. R. astr. Soc.
    175, 177.
Wright, K. O. (1952) Publ. Dom. Astrophys. Obs., Victoria, 9, 189.
Wright, K. O. (1958) Astr. J. 63, 312.
Wright, K. O. (1969) Publ. Dom. Astrophys. Obs., Victoria, 13, 301.
Wright, K. O. (1970) Vistas in Astronomy, Vol. 12, ed. A. Beer,
    p. 146, Pergamon Press, Oxford.
Wright, K. O. (1972) J. R. astr. Soc. Can. 66, 289.
Wright, K. O. and Larson, S. J. (1969) Mass Loss from Stars, ed.
    M. Hack, p. 198, Springer-Verlag, New York and D. Reidel,
    Dordrecht.
Wright, K. O. and Lee, E. K. (1956) Publ. astr. Soc. Pacific 68, 17.
Wright, K. O. and McKellar, A. (1956) Publ. astr. Soc. Pacific 68, 405.
Wright, K. O. and van Dien, E. (1949) J. R. astr. Soc. Can. 43, 15.
Wurm, K. and Struve, O. (1938) Astrophys. J. 88, 84.
Wyse, A. B. (1934) Lick Obs. Bull. 17, 37.
Zeldovich, Ya. B. and Guseynev, O. H. (1965) Astrophys. J. 144, 840.
Ziólkowski, J. (1969a) Astrophys. Space Sci. 3, 14.
Ziólkowski, J. (1969b) Mass Loss from Stars, ed. M. Hack, p. 231.
    D. Reidel, Dordrecht.
Zorec, J. (1976) preprint.

STAR INDEX

GENERAL INDEX